Commercial Metal Stud Framing

Ray Clark

Craftsman Book Company
6058 Corte Del Cedro
P.O. Box 6500
Carlsbad, CA 92018

Acknowledgments

I would like to take this opportunity to thank my family for their love, guidance, support — and for believing in me.

Mom, Julie, Uncle Richard, Johnny Boy, Margie Dale, Camilla, Jesse Ray, Amanda, Gary

And my friends and associates, who encouraged me, taught me, gave me opportunity, worked with me to bring this book into being, and most of all, accepted and befriended me when I most needed it.

Jerry & Trish Anderson, Kathy Armbruster, Dick Barker, Larry Barnes, Bob Bernsen, Rob Bigley, Jake Boyle, Hank Bomberry, Joe Buntyn, Keith Buster, Pat Cantu, John Chambers, Kelly & Angie Crow, Garland "the Kid" Curry, Steve Dechant, the people at Deming Photo Lab, Ronnie Everett, Mike Farrell, Dave Greening, John Haley, Ed Hampe, Brian Hetzel, Jim Hilton, Dwayne Kapple, Mike Kelly, Ray Kemple, Dwayne Kruger, John Landers, Gary Lowe, Jim Matt, Chris Mobley, Rick Nienke, Steve Nienke, Ol' Smoke, Terry Piece, Billy Phillips, Mark Rivera, Steve Rucker, Troy Schremmer, Terry Shaw, Rod Simms, Jack Wedge, Brett Wiggins, Don Williams, Craig Willnerd, Mike Wilson.

Dedication

This book is dedicated to the memory of my mother, Berna Mae Clark,
to Jerry Deutch, a great boss who was loved and is missed by all
and to God
for His gifts of forgiveness and perseverance.

Looking for other construction reference manuals?

Craftsman has the books to fill your needs. **Call toll-free 1-800-829-8123** or write to Craftsman Book Company, P.O. Box 6500, Carlsbad, CA 92018 for a **FREE CATALOG** of more than 100 books, including how-to manuals, annual cost books, and estimating software.
Visit our Web site: *http://www.craftsman-book.com*

Library of Congress Cataloging-in-Publication Data
Clark, Ray, 1961-
    Commercial metal stud framing / by Ray Clark.
      p. cm.
    Includes index.
    ISBN 1-57218-079-X
    1. Steel framing (Building)--Handbooks, manuals, etc. I. Title.

TH2301 .C53 1999
624.1'773--dc21   99-050302

© 1999 Craftsman Book Company

Cover design: **Emil Ihrig, Helios Productions**
Interior design: **Sybil Ihrig, Helios Productions**

Contents

Section I: Tricks of the Trade — 1

1. Wall Methods — 3
- Reading the Blueprints — 3
- The 3-4-5 Squaring Method — 4
- Laying Out the Walls — 6
- Cutting Plate and Studs by Hand — 10
- Splicing Plate — 11
- Shooting Down the Bottom Plate — 13
- Shooting Up the Top Plate — 15
- Forming a Corner with Plate — 17
- Laying Out Plate — 21
- Stuffing Studs — 22
- Slap Studs and Sliders — 23
- Notching Plate and Studs for Obstacles — 24
- Deflection Plate — 26
- Radius Plate — 28
- Fastening the Framing Members — 29
- Kickers and Other Braces — 30
- Wall Expansions — 32
- Straightedging — 33

2. Headers — 35
- Headering Door and Window Jambs — 36
- Headering Around Obstacles — 36
- Building Box Beam Headers — 37
- Stiffbacks — 39

3. Suspended Drywall Ceiling Methods — 41
- Establishing and Maintaining a Constant Elevation — 42
- Tying Suspension Wires to Bar Joist — 44
- Fastening Ceiling Materials — 50
- Tuning in the Ceiling Elevation — 53

4. Soffit Methods — 55
- Laying Out a Soffit — 55
- Figuring Soffit Stud Length — 57
- Plumbing the Corner Studs — 58
- Plating the Bottom of the Studs — 59
- Plumbing a Soffit with a Dryline — 60
- Setting the Bottom Plate to a Chalk Line — 61
- Building Jigs — 63
- Framing with Jigs — 65
- Patterns — 68
- Supporting Soffits — 70

iii

Section II: Step-by-Step Methods — 75

5 Interior Walls — 77
Walls to the Deck — 78
Freestanding Walls — 86
Chase Walls — 90
Arches — 91

6 Hollow Metal Jambs — 95
Clips — 96
Door Jambs — 97
Window Jambs — 102

7 Furred Walls — 105
Furring with DWC — 106
Furring with $1^{5}/_{8}$-Inch Studs — 109
Furring with Z-Channel — 111
Resilient Furring Channel — 112

8 Structural Stud Walls — 115
Bearing Walls — 116
Structural Stud Walls with Slide Clips — 120
Structural Stud Walls Welded to the Red Iron — 123
Parapet Walls — 123
Parapet Walls Framed to the Red Iron — 126
Diagonal Cross Bracing — 127

9 Fire-Rated Walls and Ceilings — 129
Shaft Liner Systems — 130
Fire-Rated Walls — 136
Fire Ceilings — 138

10 Columns — 141
Full-Framed Columns — 142
Framing a Cornice to the Column — 144
Speed-Framed Columns — 146

11 Soffits — 151
Spanning Speed Soffit — 152
Suspended Speed Soffit — 155
Framed Soffit — 158
Framing a Drop with Jigs — 161
Suspension Walls — 166

12 Suspended Drywall Ceiling Systems — 171
Suspended Drywall Ceilings — 172
Hard Lid Ceilings — 177
Skylights — 179

13 Drywall Methods — 183
Marking Sheets with Tape Measure and Pencil — 184
Cutting Drywall — 184
Setting an Outside Corner with a Rip — 187
Screwing Off the Drywall — 187
Tying in Slap Studs — 189

Glossary — 191
Index — 197

SECTION 1
Tricks of the Trade

Welcome to the world of metal stud framing. My name is Ray Clark and I'll be your guide to this challenging and rewarding trade. I've been working in the metal stud and drywall trades for 15 years, and have taught these trades as a Junior College instructor. While teaching, I contacted publishers from coast to coast looking for a text to use in my class. I couldn't find one. So I started writing one for my class — and it evolved into this book.

To keep the information simple, to the point, and useful for both journeyman wood carpenters and apprentices alike, I've organized the book into two sections. The first, Tricks of the Trade, focuses on the unique methods and techniques common to the metal framing trade. The second, the Step-by-Step Methods section, concentrates on the process and flow of work involved in framing metal stud walls, ceilings and soffits. As we work through each chapter, I'll also introduce you to tools that are common to the trade, but which may be new to you.

We'll cover the many advantages of framing with metal studs as opposed to the traditional wood. The most obvious is that metal studs won't burn, rot or become a termite buffet. Structurally, as in bearing walls, metal studs are many times stronger than wood studs, greatly reducing the amount of materials needed to support the load, as well as the amount of time it takes to frame the job. And the fact that you screw metal studs together as opposed to nailing wood studs also makes them stronger and faster to assemble. The metal stud material doesn't dry out, either, so warping, bowing and twisting aren't problems. That lets you build straighter walls and flatter ceilings.

And here's another advantage. Wood prices are jumping all over the place. You can bid a job in January, based on lumber prices at that time, then get a nasty surprise when you get the job and order the lumber in February. Maybe you'll find an owner who's sympathetic and understanding about your request for more money. But probably not. I never have. Steel, on the other hand, remains pretty stable. Weather, politics, overharvesting, environmental issues, rarely come into it. Plus, you don't have to worry about quality. Twenty gauge steel is 20 gauge steel, no matter where it comes from. With wood, especially lately, you never know what you'll get.

Commercial metal stud projects range from small, tenant finish jobs that'll only take a few hours to frame up, to extremely large project that often last for a year or more. Most metal stud contractors, whether union or not, will pay by the hour, with wages ranging from a low of $7 in some areas of the country to $20-plus, depending on your experience, in others. Many contractors also offer benefit packages including health insurance and 401(k) plans, as well as paid vacations and holidays.

In this section we'll discuss some commonly-used tricks of the trade in metal stud construction. They're shortcuts and methods that have become standards of this trade. Obviously, knowing the tricks of the trade gives you an edge. First, it'll reduce the time it takes you to reach journeyman status. Second, you'll know "what's going on" when you begin working with a new partner or new outfit. In either case, the result is the same. It'll help you become more proficient, which makes you a better hand, which means you earn more money. Isn't that the reason you bought this book?

Consider this section as a reference guide that lets us cover the principles of metal stud construction without getting sidetracked on the details of how to accomplish each particular step. This is general knowledge you need under your belt before you actually begin putting up metal stud walls. If you're already an experienced wood carpenter, you may know some of this already. If that's the case, skim through any familiar material in this first section. But I don't recommend skipping anything entirely. You never know when you might pick up a new idea or improved method. It's probably worth your time to read these first four chapters just in case there's a trick or technique that's new for you. And I'll bet there is. Nobody knows it all — not even the authors of books about it.

As you put these methods to work, experience will quickly teach you in where you can put them to work. You can also use this section as a reference guide in connection with the step-by-step directions in the second section of the book.

Chapter 1

Wall Methods

Reading the Blueprints

Reading blueprints and layout are subjects too large for us to cover thoroughly here. There are many books available about both subjects. I'll only cover the fundamentals of layout work and blueprint reading to give you a basic understanding. Experience will teach you much more. If you need more information, look at the order form in the back of the book for *Blueprint Reading for the Building Trades* and *Building Layout*. In this chapter, I'll just cover the basics of print notations.

The *magnetic north arrow,* shown in Figure 1-1, is located on the right-hand side of each page of the prints. It helps keep all of the work on the job site going the same direction.

The *detail symbol* indicates a specialty item or condition in a wall, and gives the location of a detailed drawing for the item. The detail symbol in Figure 1-2 refers the reader to detail A (top letter) on page A2.4 (bottom number). The detailed drawing is often called *a cut.*

The *wall legend,* also known as a *key* (Figure 1-3), distinguishes the various wall types in the prints. Each wall type in the legend

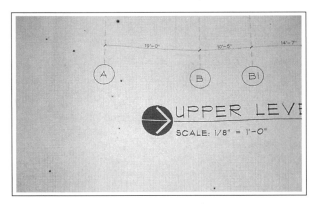

Figure 1–1. This shows a north or "magnetic north" arrow which points to true north, and helps keep everything going in the same direction.

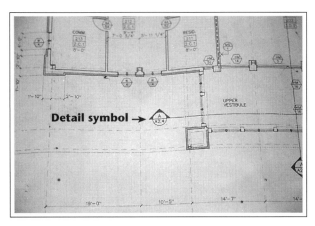

Figure 1–2. This detail symbol is directing the carpenter to a detail (or cut) "A" on page A2.4 in the prints.

has a detailed description and a number. The wall numbers in the legend coincide with wall numbers in the floor plan. The wall legend tells you what type of metal stud framing material you'll use for each wall. It also lets you know whether the wall is freestanding or framed to the deck. If it's a freestanding wall, it also gives you the required height. The wall legend also indicates the thickness, type and the number of layers of drywall used, and any insulation materials.

The *reflective ceiling section* of the prints (Figure 1–4) gives all the ceiling elevations and the material they're to be built of. Rooms that show light grid lines have a grid ceiling, while clear rooms have a drywall ceiling. The dimensions and elevations for soffits are also given in the reflective ceiling plans, as well as the location of recessed lights and HVAC vents in the ceiling. All elevations are finish elevations, so you've got to add the thickness of the drywall to achieve the frame line elevation.

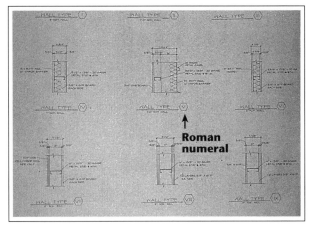

Figure 1–3. This wall legend uses Roman numerals to distinguish the different types of walls on the job.

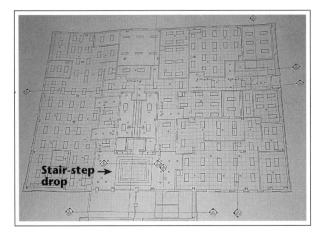

Figure 1–4. The reflective ceiling plan shown here breaks this floor of the building into "grid ceilings" and "hard lids." You can also see a large stair-step drop in the entry way in the rectangle grid near the bottom of the plan. The lines around the rectangle show the separate widths of the stair-steps. The detail symbol cutting through the drop will give the rest of the information.

The 3-4-5 Squaring Method

Let's begin with one of the most basic tricks in construction—making a right angle that's exactly 90 degrees. If you can't do that, you'll create problems that will affect not only your work, but that of all the trades that follow you.

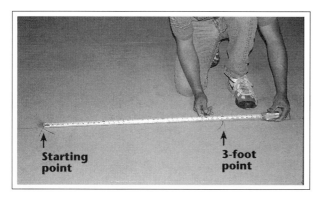

Figure 1–5. The squaring method begins with marking first the starting point on the chalk line, then the 3-foot point.

Figure 1–6. Measuring from the 3-foot mark and striking an arc at 4 feet, using the same edge of the tape at both points.

The 3-4-5 method is a simple way to square a perpendicular line off an established wall or reference line using only your tape and pencil. It's accurate, it's easy and it's faster than setting up a transit or laser to do the same job. There are five steps in this method:

Step 1 Mark a crow's-foot anywhere along the reference line (see Figure 1–5). That's your starting point for the squaring process.

Step 2 From the first crow's-foot, measure straight down the reference line 3 feet and mark a second crow's-foot.

Step 3 From the 3-foot mark, measure 4 feet off the reference line, as close to 90 degrees as possible, and strike an arc approximately 1 foot long (shown in Figure 1–6). As you draw the arc, hold the tape measure to the crow's-foot on one specific edge of the tape. To ensure accuracy, you've got to hold the pencil on the same edge of the tape while striking the arc. Striking the arc is easiest as a two-man job. But if there's no help close by, drive a concrete pin into the pivot point and hook the end of your tape to the pin.

Step 4 Return the end of your tape measure to the first crow's-foot marked on the reference line, and from there strike a second, intersecting arc at 5 feet (see Figure 1–7).

Step 5 Next, pull a chalk line from the 3-foot crow's-foot on the reference line, through the point where the two arcs intersect (Figure 1–8). Pull the chalk line quite a ways past the intersecting arcs.

Figure 1–7. With the end of the tape anchored at the starting point, strike the 5-foot arc intersecting with the 4-foot arc.

Figure 1–8. Pulling the chalk line from the 3-foot mark through the intersecting arcs completes the 3-4-5 squaring method.

Just be sure you can see the chalk line pulling through the intersecting arcs. To square larger areas, double the 3-4-5 measurements to 6-8-10.

Laying Out the Walls

Establishing the reference lines and laying out the work area are the first steps in metal stud framing. The layout work is critical to a quality frame job. That's why it's entrusted to only the top hands on a job. The reference line (sometimes called the *gospel line*) is the centerline of the job. All the other wall lines will be established from the gospel line. It *has* to be right.

Establish the gospel line by measuring the overall width of the concrete slab (or pad) at the two opposite ends of the building. Or you can establish the gospel line from the red iron columns to the structural steel. Then mark half of the overall width at each end (Figure 1-9). Next, snap a chalk line from mark to mark. For long slabs, use a laser to make multiple center reference marks so you can snap a consistently straight chalk line. Finally, spray clear enamel over the chalk line to protect it. You want it to last until the layout is complete.

With the gospel line in place, your next move is to establish a perpendicular reference line exactly 90 degrees to the gospel line. You can establish it with a transit or laser, or with the 3-4-5 or 6-8-10 method (as long as you do it carefully and accurately).

Now check the floor plan section of the blueprints to determine the layout of the wall line. Begin with the exterior walls and then move to the interior, starting on the long walls first. The hallways are a good starting point, for two reasons:

1. They're long continuous walls you can use to establish other parallel wall lines.
2. The hallways are among the few walls on a job that have very little tolerance for variance. The *Accessibility for the Disabled Act* requires you to meet stringent guidelines for width. Bathrooms are another area where wheelchair accessibility is very important, so you don't have much tolerance for these walls either.

As you lay out the wall lines, remember to allow for the thickness of the drywall. Forget this and your walls will be more than an inch short. When the building inspector takes out his tape measure, you're done for! All wall line dimensions in the prints are *finished walls* unless otherwise noted. You have to consider all the thickness of all the layers of drywall. You'll find this information in the *wall legend* section of the prints.

After you've marked the starting walls (exterior walls and hallways), work your way through the job from end to end, snapping all the long walls first. As you figure the wall dimensions, mark them as close to the ends of the wall as possible. Then snap a chalk line between the marks. It's common to add the width of the framing material and snap a line for both sides of the bottom plate (Figure 1-10). This eliminates a common mistake: plating the wrong side of a wall line. Snapping a chalk line to both sides of a wall also makes it much easier to mark the next wall, since you measure it from the wall you just completed. If you don't snap both

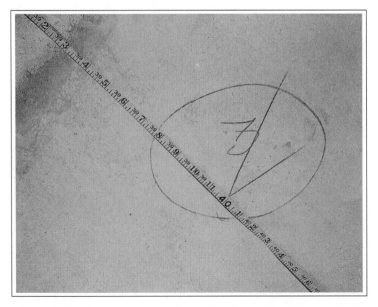

Figure 1-9. Here you can see the center of an 80-foot slab marked near one end of the pad.

sides of the plate, always mark an "X" on the side of the line where the wall will sit (Figure 1-11).

Laying Out the Doors and Windows

As you're laying out the walls, it's important to lay out doors and windows at the same time, completing each wall as you go. Square the wall lines off on both sides of the door opening and clearly mark the opening DOOR. Then write the door number inside the opening and mark the swing of the door (Figure 1-12). Lay out the opening for a door 4 inches wider than the door's width according to the prints. The dimensions in the door schedule section of the prints are the size of the door itself. You need to add 4 inches (2 inches on each side) for the jamb.

After you've chalked the wall lines, go back and lay out the windows. Write the window number between the layout marks, as well as the elevation of the bottom of the jamb. Figure 1-13 shows a properly laid-out window. The window studs are clearly marked right at the wall line. The information (42" off FF 40" x 40") tells the framer that the bottom of the window jamb is 42 inches off of the finish floor, and the window is 40 inches tall and 40 inches wide.

Every outfit has its own customs, so you won't find windows laid out like this on every job. It's also an accepted practice to lay out the window jambs *after* the plate is shot down. In this case, the layout for the window studs is marked along the edge of the plate with the layout for the studs. You can also use this method for laying out items like fire extinguishers and tissue dispensers that are recessed in the wall. Write their elevations inside the plate between the stud layout marks. You'll find items recessed in a wall noted in the prints on the wall line with a *detail symbol*.

It's important to maintain consistency in the window and door jamb elevations. It's common to find an unlevel or poorly-floated pad that causes the top elevation of the jambs to be uneven. In most cases, the jambs should have a common elevation throughout a given work area. If they don't, it'll be noticeable when the room's finished—and then it's too late. Make bench marks with a water level or transit, or use a laser to make sure they're consistent.

Layout Around an Obstacle

What if there's an obstacle that prevents you from snapping a chalk line on the wall line? Here's a four-step method you can use to extend an accurate chalk line around any obstructions.

Step 1 From a reference or wall line, measure out and mark two crow's-foot marks as far apart as possible.

Figure 1-10. In this example you see an intersecting 3⅝-inch wall line laid out with both sides of the wall snapped.

Figure 1-11. One side of a wall line snapped out with X's marking the side of the line the wall will sit on.

Figure 1-12. A door laid out along the wall line with the door number and swing marked out in the door opening.

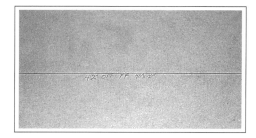

Figure 1-13. The rough opening for a window laid out along the wall line, telling the carpenter that the window R.O. is 42 inches off the finish floor and 40 x 40 inches.

Step 2 Hold the end of the chalk line down to the first crow's-foot, then pull the chalk line past the second crow's-foot, as far as you can accurately line up the chalk line on the crow's-foot (Figure 1–14).

Step 3 While you hold the chalk line tightly to the first crow's-foot, your partner will wrap the opposite end of the chalk line around a flat carpenter's pencil to keep it raised slightly off of the floor (Figure 1–15). Then your partner can adjust the chalk line until it's positioned directly over the second crow's-foot (Figure 1–16). If your partner is so far away from the second crow's-foot that he can't accurately set the chalk line to the crow's-foot, you'll need a third person to help. The third person will also hold the chalk line down to the floor in the center after it's adjusted into place, and snap the chalk line on each side.

Step 4 Check the line just snapped for accuracy by looking to make sure that the chalk line came exactly through the crow's-foot. If the chalk line is off, even slightly, erase it and start over again with a different color of chalk. If the line is off by $1/4$ inch in 10 feet, it'll be off by $1/2$ inch in 20 feet. It's surprising what trouble a $1/2$ inch can make.

Plumbing with the Plumb Bob

The plumb bob is a very simple, accurate and quick way to transfer a wall line from the floor to the deck. Make sure the plumb bob has a string line that's braided, not wound. That helps prevent the plumb bob from spinning excessively. Here are some common techniques for plumbing up.

The first step in the plumbing process is to adjust the elevation of the plumb bob, so the tip is between $1/8$ and $3/8$ inch above the floor or edge of the plate. A quick way to achieve the desired elevation is to lace the plumb bob string through the fingers of your hand holding the plumb bob. Run the string line over your little finger, then under the ring and middle fingers, and over the index finger to the thumb (Figure 1–17). The thumb will hold the string line tight to the deck or other surface you're plumbing to. When you release pressure on the thumb, it's easy to raise and lower the plumb bob, with control.

Figure 1–14. In this example you see the chalk line being pulled well past the far crow's-foot, and adjusted to the crow's-foot.

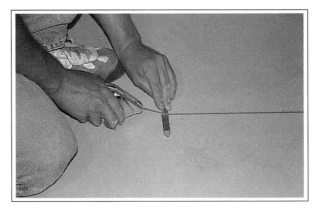

Figure 1–15. Wrapping the chalk line around your pencil not only keeps the chalk line off the floor as it's set to the crow's-foot, it will also keep the line straight over humps in the slab.

Figure 1–16. Here the chalk line has been set and is ready to snap.

Chapter 1: Wall Methods   9

Figure 1–17. The string line of a plumb bob properly run through the fingers and held to the deck with the thumb. Notice that both the wall line plumb point and a 90-degree plumb mark were marked.

The tip of the plumb bob must be steady to get an accurate reading. To steady the tip, your partner will place both hands close together on the floor, one hand on each side of the plumb bob. By raising the index finger of each hand to the tip, with light and equal pressure, he can steady the plumb bob (Figure 1–18).

In directing the movement of the plumb bob, use clear concise terms like "left an eighth" or "right a quarter." These directions are quicker and easier to understand when they're coupled with hand signals. Once plumb is achieved, sing out loud and clear. Don't leave your partner guessing about what's going on.

On the deck, make two crow's-foot marks at the exact point of the string line, one mark in the same direction as the wall line below, the second mark at 90 degrees to the wall line. You can see these marks in Figure 1–17. Also mark the 90-degree point at the bottom plate or wall line. The 90-degree plumb marks will be used later as a reference mark. From this reference mark, transfer the layout marks for the studs from the bottom plate to the top plate. You can also use the plumb marks to establish perpendicular wall lines by measuring from the reference mark to the perpendicular wall line. Then transfer the measurement to the deck above (Figure 1–19).

The pocket laser is a great advance in carpentry tools. These battery-powered lasers are accurate to about 50 feet, and allow one person to do the plumbing work. Bump the laser up to the plate and measure from the laser beam 1 inch to the plate line, as shown in Figure 1–20. The laser can also be set right to the wall line,

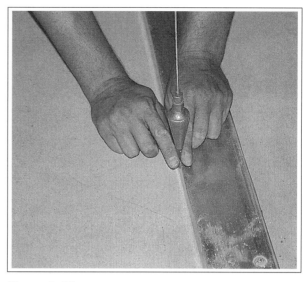

Figure 1–18. Steadying the tip of the plumb bob with light, even pressure from each index finger.

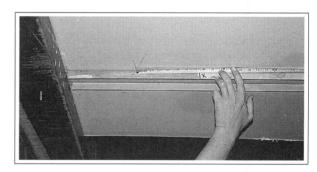

Figure 1–19. Measuring from the 90-degree plumb mark to establish the stud layout at the top plate.

Figure 1–20. A pocket laser set up against the bottom plate of a $3^{5}/_{8}$-inch wall. While the wind and vibration will affect the pocket laser to a certain extent, it's still much faster and easier to use than a plumb bob. Use caution, though; never look directly into the beam. It can damage your eyes.

matching the hash marks on the laser to the chalk line. These little lasers also shoot a level beam, and both the plumb and level beams form a square 90-degree angle.

Cutting Plate and Studs by Hand

There are two ways to cut your metal plates and studs: with snips or by scoring them. We'll start by cutting with snips.

Cutting with Snips

To begin with, let me make a couple of recommendations. First, *always* wear leather gloves while working with metal stud material of any kind. (The carpenters in our examples aren't wearing gloves to give you a clear view of the work being done.) Second, if you're going to start with just one pair of the snips used in the trade, I recommend the straight cuts. The snips are available in three different directional cuts. Red-handled snips cut left, green-handled snips cut right, and yellow-handled snips cut straight. You'll have to choose the snips that are most comfortable for you. Eventually, you'll probably want to have all three cuts.

Cut light gauge (25, 22 and 20 gauge) studs and plate in two smooth fluid moves. But this smooth fluid movement comes with experience. We'll break the cutting process into three steps:

Step 1 Measure out and mark the length of the cut, marking the cut with your snips instead of your pencil (Figure 1–21). This eliminates the time spent switching the pencil for the snips. But be careful not to cut your tape measure—especially if someone's watching!

Step 2 Square the cut from one leg, across to the other leg of the material. Whenever possible, eyeball the cut across from one leg to the other. Eyeballing cuts is an important skill that you need to master as quickly as possible. Sight through the cut in the first leg of the stud to the other leg, and cut it with your snips (Figure 1–22). Mastering this step takes practice. Use some scrap stud to work on until you can make the two cuts and fold the material back over with the edges lining up evenly. The stud in Figure 1–23 has a perfect square cut. While this is the ideal, there's usually some room for error. If the situation calls for a perfectly-square cut, use a speed or combination square to mark the cut (Figure 1–24).

Step 3 Fold the material back at the cuts you made in the two legs, then unfold the material. Using your free hand, grab one side of the material 2 to 3 inches from the

Figure 1–21. Marking the length of the cut with snips.

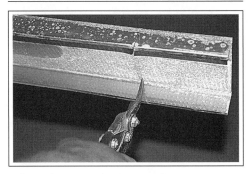

Figure 1–22. Eyeballing across the stud from the first cut to the second.

Figure 1–23. A stud cut squarely, using snips. You can tell if the cut is square by folding the stud back over and making sure the edges of the stud line up.

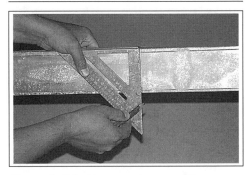

Figure 1–24. The cut point squared across the hard side of a stud with a speed square, with one leg of the stud cut to the squared line.

cut. Next, cut across the fold line in the material. As you're making the cut, grip the material tightly so it rolls out of the way of the cut (Figure 1–25). Rolling the side of the stud out of the way will allow the snips to pass through your material and complete the cut.

Cutting by Scoring

Use this method to cut heavy gauge material using only hand tools. It's particularly useful when there's no chop saw available, or it's in use. The procedure is similar to cutting with snips, so I won't go into great detail.

Step 1 After marking the length of the cut on the hard side of the stud in pencil, set your speed or combination square to the mark. Using your utility knife, score the hard side of the stud, two to four times depending on the sharpness of the knife blade. Make certain that your hand is above the direction of the cut, and that your body is positioned to the side of the cut (Figure 1–26).

Step 2 Using your snips, cut the legs of the stud at the score. Heavy gauge bullnose cutters are the ideal snips for cutting structural material, or keep an old pair of snips around for this purpose. That prevents your good snips from being trashed.

Step 3 After completing the cut, fold the stud back and forth a few times until it breaks along the score.

Splicing Plate

Splicing plate (or *track,* as it's also called) is the common way of joining the sticks of plate to form one long continuous piece. You'll have to splice the plate in many situations, including the top and bottom plates of walls, and in suspension systems like soffits and suspension walls, to name just a few. Unless you're framing in an expansion, everything has to tie together. Tying the metal stud framing together provides much of its strength. I'll break the process into steps to describe it, but when you do it in the field, it'll become a fluid movement.

1. Using your snips, cut a 1-inch to 1$^1/_2$-inch-deep slice in the hard side (back) of the plate, as shown in Figure 1–27. Keep in mind that the cut for the splice is always made in the next piece of plate to be set in place.
2. Next, place the splice against the end of the plate already shot in place. Figure 1–28 shows two sticks of

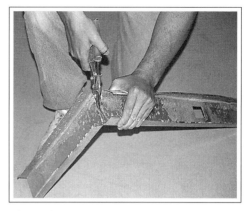

Figure 1–25. The scrap side of the stud rolls up out of the way if you grip it tightly, allowing the snips to easily pass through the cut.

Figure 1–26. Using a speed square to "score" a 6-inch structural stud with a utility knife. The stud sometimes needs to be scored several times. The better it's scored, the quicker it will snap off when you fold it back and forth.

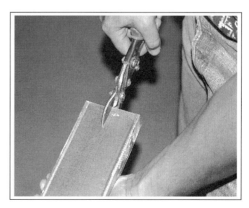

Figure 1–27. Cutting the slice in one end of a stick of 3$^5/_8$-inch plate, using snips.

bottom plate being spliced correctly. Leave the end of the first stick loose so you can slide the two pieces together. The splice also acts as a third hand to hold up one end of the plate while you slip it onto the studs in suspension work. Figure 1-29 shows this happening while plating the top of a freestanding wall. Any time you're plating the bottom of the studs in a soffit or suspension work, clamp the splice together. In Figure 1-30, the plate is clamped right in the corners to prevent any offsets in the plate — either in and out or up and down. Tack the splice together with only one framing screw, then add more screws to the other side and bottom of the plate as you set it to the string line. This prevents kinks from forming at the joints of the plate.

3. In walls running from floor to deck, hold the spliced joints cut into the top and bottom plate to the wall line and shoot them in place (Figure 1-31). As you can see, the ends of the plate are spliced and then shot down with one pin. This technique creates one continuous length of plate while saving one pin and load per joint.

Splicing or Scabbing Structural Plate

To splice structural plate, follow the same steps, except you'll make the splice with a chop saw or your old pair of snips. (Making cuts on structural material with your good pair of snips will make them an old pair real quick.) The problem you'll have with splicing the heavy gauge material is that the material's thickness may cause a bump in the finish material. In some situations you can get by with this, but in others you can't. The type of finish material will dictate your splice. When a splice won't work for you, scab the plate together. A scab is simply a scrap piece of stud cut from the same width material as you're using to frame the wall. Scab the plate together following these steps:

1. First, cut the scab about 6 inches long, so each piece of plate has roughly 3 inches of the scab to screw into. Slide the scab into the first stick of plate already in place, and clamp in the corners. Then fasten the scab

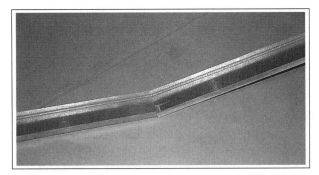

Figure 1-28. Two sticks of bottom plate being spliced together.

Figure 1-29. The splice will hold up one end of the top plate in this freestanding wall.

Figure 1-30. A joint in the plate is clamped directly in the corners of the plate.

Figure 1-31. The ends of two sticks of bottom plate spliced together and shot down with one pin.

Figure 1–32. Here a scab is clamped and screwed off into the first stick of 6-inch structural plate, forming the joint. Wider 6-inch deflection plate was used in this example to highlight the scab itself.

in place with one self-drilling framing screw (commonly referred to as an S-12 or pan head), as shown in Figure 1–32.

2. Now, slide the next piece of plate onto the scab as it's slipped onto the studs. Then clamp and fasten it in place with one S-12. The two screws will hold everything together until you straighten the wall with a string line. Then run three more screws in each side of the scab.

Shooting Down the Bottom Plate

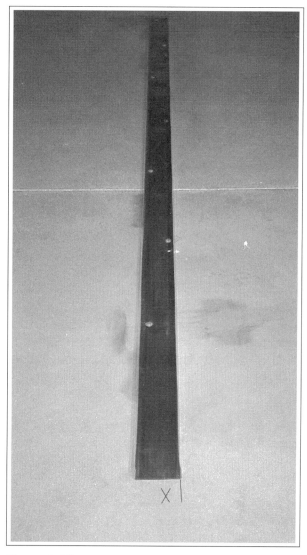

Figure 1–33. Bottom plate shot down following the wall line with the plate sitting on the side of the wall line marked with an X.

As you shoot down the bottom plate (fasten it in place with a powder-actuated nail set), it must follow the wall line precisely. Plate that doesn't follow the wall line will create dips and bumps in a wall that may make the wall unacceptable. In situations where only one side of the wall line has been snapped, shoot the plate down on the side of the wall line marked with an "X," as shown in Figure 1–33.

You'll splice light gauge plate at the joints as you shoot it down. With heavy gauge plate, you can simply butt the ends together at the joint (Figure 1–34). Space the pins approximately 24 inches apart in either case, as shown in Figure 1–35.

Figure 1–34. Bottom plate on a 6-inch structural stud wall butted together and shot down.

Commercial Metal Stud Framing

The bottom plate runs continuously under windows and other wall penetrations. As you shoot it down, transfer any layout information for these items from the wall line to the plate with a black felt tip marker. At door jambs and wrapped openings, stop the bottom plate $1/8$ inch short of the opening (Figure 1-36). The exception to this rule is when you're installing a computer floor. In that case, the bottom plate runs through the door jamb's rough opening. We'll look at this situation in detail in Chapter 7.

Sometimes the bottom plate will be interrupted by obstacles in a wall line, such as plumbing lines and large conduit. In the plumbing walls (chase walls) of bathrooms, for example, you may have to piece in the bottom plate (Figure 1-37).

As you shoot down the bottom plate, establish the sliders (or slap studs) at intersecting walls and inside corners. Frame the slider into the shorter walls, so that they tie into the longer wall of the corner. That's important because it allows a $3/4$-inch gap between the two pieces of plate that form a corner in the metal stud walls. The gap allows the drywall to slide inside the corner. Then, as the walls are rocked, the slider is tied into the drywall to form a solid corner. There are two common types of corners:

1. The *inside/outside corner* is shown properly plated in Figure 1-38. This corner is a widely-accepted method of plating an inside/outside corner.

2. The *double inside corner,* formed by a partition wall tying into a long wall (such as a hallway), is shown properly plated in Figure 1-39.

Shooting down the bottom plate is a two-man job. As one carpenter makes any needed cuts, including splicing the plate and putting it in place along the wall line, his partner follows behind setting the plate to the wall line and shooting it in place.

While you're shooting the plate in place, there'll be some changes or mistakes that'll make it necessary to pull the plate back up. To get the pins loose quickly and easily, you can shoot them loose with the shotgun. Using the shotgun with no pin in it, set the barrel right on top of the pin you want to pop loose, and fire. The piston in the shotgun will drive the pin on

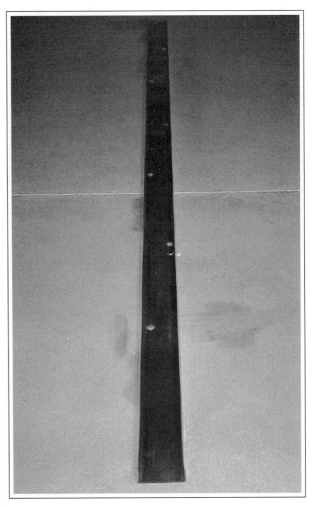

Figure 1-35. You can see the first stick of bottom plate shot down for a $3^{5}/8$-inch wall with the pins staggered from side to side every 24 inches. Notice the end is left loose so the next stick can be spliced to it.

Figure 1-36. The bottom plate of a $3^{5}/8$-inch wall stopped for both a corner and a door.

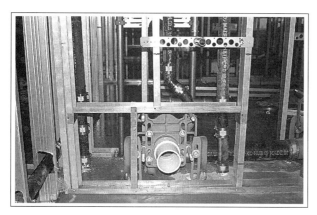

Figure 1–37. Here's a good example of a congested plumbing wall that required a header at the bottom of the wall to get around an obstacle.

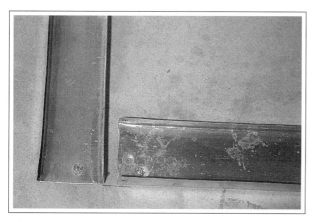

Figure 1–38. Here we see the bottom plate of a 3⁵⁄₈-inch wall forming an inside/outside corner, leaving a ³⁄₄-inch gap for the drywall to slide inside the wall.

Figure 1–39. A ³⁄₄-inch gap left open between the bottom plate of two intersecting walls, forming a double inside corner.

through the plate into the concrete. Never use this technique when the pins are shot into steel. Instead, grab the pin shot into the steel with your end nippers and work the pin back and forth until it comes loose.

Shooting Up the Top Plate

When you start shooting up the top plate, you might expect that you could simply learn a few standard procedures and get to work. That's true—up to a point. The standard procedures will carry you through the "gravy" work. But you need more than that. We'll cover the basic principles of plating the deck, *and* the deck conditions you're likely to find which will complicate the process. You need to know what problems to look for, and how to overcome them.

The Basics

The basic methods for plating the deck are a lot like those for shooting down the bottom plate. Shoot the plate to the deck with a gas- or power-actuated nail set, following the wall line exactly and spacing the pins approximately 24 inches apart. The splice plays an additional role with the top plate. You can use the splice to hold up one end of a stick of plate as you fasten the opposite end in place.

The top plate follows the bottom plate's slider (slap stud) placement, with the same ³⁄₄-inch gap (Figure 1–40) to allow the drywall to slide inside the intersecting wall at inside corners. The outside corners are also determined by the bottom plate, using the 90-degree plumb mark to establish the corner points and the stud layout for the wall. Run the top plate continuous unless there are obstacles that prevent it. Stop the top plate at all wall expansions, leaving a ³⁄₄-inch void in the plate. Plumb the wall expansion from floor to deck.

Concrete Decking

Concrete decks are typical in multistory buildings. They're formed of raw concrete, corrugated metal decking with concrete poured on top of it, or a prestressed concrete truss system (Figure 1–41). You can fasten the plate to any of these

Commercial Metal Stud Framing

decks with a gas- or powder-actuated nail set. But don't shoot into corrugated metal decking until you've checked that the concrete has been poured. If you shoot a pin to unpoured metal decking, it won't hold the plate in place. It also poses a serious safety hazard. In this situation, screw the plate to the deck using S-12s. If the deck is concrete trusses, you may have to span them with studs that have tabs cut on the ends. Shoot them to the bottom of the ribs, then screw the top plate to the studs.

Metal Decking

In a single-story commercial building like department stores and malls, corrugated metal decking is commonly used as the roofing material. Screw the top plate to the deck with S-12s. But there are two common problems.

First, the long decking screws used to fasten down the roofing materials often come through the decking in the wall line. They're in the way of the top plate. To solve the problem, snap off the screw close to the deck (Figure 1-42). Simply grab the screw close to the deck with your lineman's pliers and bend it until it breaks off.

Second, the top plate may line up inside the concave groove of the corrugated metal decking. That makes it difficult to attach the plate to the deck, as well as to fasten the studs into the plate later. To solve the problem, cut short pieces of plate with tabs and screw them to the deck spanning the concave groove. Space the bridging 24 inches on center and screw it to the decking with S-12s, or shoot it up if the decking has concrete poured on top. Once in place, plumb and snap the wall line to the bridging. Then fasten the top plate to the bridging with S-12s or tek screws (Figure 1-43).

Plating to the Bar Joist

In some framing situations, it's acceptable to attach the top plate to the bottom of the bar joist (a structural steel roofing and flooring joist). See Figure 1-44. There are two advantages to doing this. First, it eliminates the need to plumb up the wall line between each set of bar joists. Second, because the studs will only run up to the bottom of the bar joist.

Figure 1–40. Here you can see the top plate fastened together using the tab method. The plate is cut, leaving a 3/4-inch gap to allow the drywall to pass through the wall.

Figure 1–41. This prestressed concrete truss system supports the floor above.

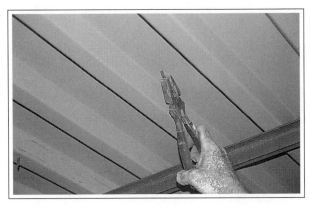

Figure 1–42. Grab a long decking screw near its base with lineman's pliers and bend it over, snapping it off.

The type of fastener you'll use to fasten the plate to the bar joist depends on the bar joist's hardness. In some cases you can use S-12s, as long as they take to the bar joist without stripping out. If the bar joist is too hard to use S-12s, use a powder-actuated nail set to fasten the plate in place. When you use a shotgun to fasten the plate to a bar joist or any part of the red iron superstructure, always use steel pins. Shoot the pins into the bar joist close to the angle (inside edge) in the steel (Figure 1–45). Shooting the pins near the outside edge will cause the bar joist to bend under the force of the shotgun.

While you're plating the bar joist, use clamps to hold the plate in place on the wall line as you're shooting it to the bar joist. That prevents the plate from slipping — or even falling — as it's shot to the bar joist. You can avoid having to drop a plumb bob to establish the outside corner by using the 90-degree plumb mark we discussed earlier. Outside corners will often fall between two bar joists. To make a strong corner, run the top plate on past the actual corner point to the next bar joist, then fasten it as shown in Figure 1–46. Later in the chapter I'll cover how to form the corners.

When you're framing a fire, smoke or sound wall to the bottom of the bar joist, it must continue to the deck. As you stuff the studs (install them into the plate), cut scrap stud or plate to fit in between the bar joist, then fasten it to the deck. To continue the wall line up to the deck, set a straightedge, either a drywall rip or a level, to a wall stud and slide it up to the deck. Then mark the wall line, as shown in Figure 1–47.

Forming a Corner with Plate

There are two basic skills every metal stud framer needs:

1. The ability to determine the type of corner you'll need in a given situation
2. The ability to quickly and correctly cut the plate for the corner you've chosen

Figure 1–43. The concave grooves in a corrugated metal deck spanned with tabbed plate. The top plate of a $3^{5/8}$-inch wall is screwed in place to the spreaders.

Figure 1–44. The top plate of a $3^{5/8}$-inch wall screwed to the bottom of the bar joist using S-12s (self-drilling framing screws).

Figure 1–45. Here we see a pin shot right at the bend point of the angle iron at the bottom of the bar joist.

Figure 1–46. The top plate of a 3⁵⁄₈-inch wall run past the actual corner point to the next bar joist and shot in place, with the top plate of the intersecting wall tabbed and set in place.

Figure 1–47. Using a 4-foot magnetic level to continue the wall line on up to the deck. If you look closely you can see the bottom of a stud up on the floor above.

Here we'll take a look at four common methods of forming a corner with both plate and utility angle. Each style of corner is useful in certain framing situations.

Top Plate Outside Corners

To form this type of corner, begin with the first of the two intersecting plates fastened in place along the wall line, with the end roughly $1/4$ inch short of the actual corner point. As you form the corner, notch the first stick of plate for the intersecting wall that'll form the corner. Measure back from the end of the plate a distance equal to the width of the material you're using, plus $1/2$ inch. Cut the inside leg of the plate with your snips and fold the cut material up out of the way (Figure 1–48).

Now set the end of the plate up to the corner, overlapping the end of the plate already fastened in place. That holds up one end of the plate while you clamp the opposite end in place to the wall line. Eyeball the end of the plate to within $1/4$ inch of the actual corner point (Figure 1–49). Make any needed adjustments, then fasten the plate in place. In freestanding or suspended conditions, screw the corner together using three framing screws. You'll find this technique useful when framing to bar joist and freestanding walls, and also when framing to the deck. You'll also use it for capping jigs and parapet walls.

Intersecting Walls

There are two common types of corners for intersecting walls, depending on whether the two walls have equal plate elevations.

Walls with Equal Plate Elevation

You'll use this type of corner for framing an intersecting wall into a long wall that runs past, such as a partition wall to a hallway wall. To form this corner, begin by cutting about a 2-inch tab on one end of a stick of plate. Now set the tab on top of the plate of the wall that's already in place. The tab will support one end of the plate as you position it on the wall line at the other end and stand up a stud to support the top plate. Don't forget to leave a $3/4$-inch gap for the drywall to slide through. When the plate's in position, clamp it in place, then fasten it with the appropriate fasteners, as shown in Figure 1–50. This style of corner will work in most situations where the top plates of the two intersecting walls are at the same elevation.

Walls with Offset Plate Elevations

Use this corner when you're tying together two walls with different plate elevations. After the long wall is framed, figure out whether the new wall line will hit a stud or fall between two studs. If it hits a stud, first run a diagonal brace

Chapter 1: Wall Methods   19

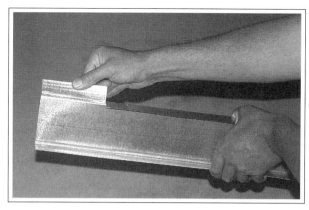

Figure 1–48. The inside leg of the plate is cut and folded out to allow it to overlap.

Figure 1–49. You can see the plate set in place overlapping the first stick of plate to form the corner.

Figure 1–50. The top plate of an intersecting wall tabbed and screwed in place with four tek screws. The shoe of a kicker is screwed in place right next to the corner joint.

Figure 1–51. A shoe cut on one end of the top plate of a shorter intersecting wall is screwed off to a layout stud of a taller wall.

Figure 1–52. Here the top plate of the shorter intersecting wall is tabbed and screwed off to a spreader.

from the top plate to the stud you're going to fasten the top plate to. With the stud plumbed and braced, simply cut a shoe on one end of the plate, and clamp it to the stud at the desired height (Figure 1–51). Slip the opposite end of the plate onto a wall stud that's been screwed off on layout in the bottom plate. This will carry the top plate as you work the wall.

When the wall line falls between two of the layout studs, cut a spreader from scrap stud material with tabs on each end. Use it to span the gap between the studs, using the tabs to fasten the spreader to the wall studs (Figure 1–52). To figure the spreader length, measure between the studs at the bottom plate, minus an additional 3/8 inch for play. Put the spreader 3/8 inch higher than the length of the studs used in the wall, and screw it to the wall studs. As you fasten the spreader to the studs, make sure you maintain the 16- or 24-inch stud layout. A temporary diagonal brace from the top plate of the original wall, down to just over the spreader, will hold the new wall plumb (Figure 1-53).

Screw the brace to the top plate and clamp it to the studs over the spreader until you get everything plumb. Here's how to do it. Measure from one of the studs supporting the spreader to the bottom plate of the wall you're framing. From the same stud, measure over and mark the same number on the spreader. Put an "X" on the side of the mark that you'll set the plate to. Now clamp the top plate of your wall to the bottom of the spreader with the edge of the plate on your mark. Just tack the plate in place with one framing screw for now; you'll add another after you've plumbed the other end of the plate. Now the diagonal brace will hold everything plumb when you plumb up the top plate, because the spreader holds the stud and plate. When the tall wall is rocked, tie the slider of the short wall into the drywall and remove the diagonal brace. This technique works well for framing a freestanding wall into a wall that's framed up to the deck.

Tying Into Preexisting Conditions

To form this type of corner, fasten the first stud of the wall to a preexisting condition, whether it's drywall, concrete, steel or a block wall. Naturally, you'll want to plumb the stud with your level while you're fastening it. In this case, simply screw the top plate to the end stud, securing the corner at the top plate (Figure 1-54). An alternative method is to use a shoe at one end of the plate. Flatten out the shoe and fasten it to the preexisting condition $1/4$ to $3/8$ inch higher than the studs used to frame the wall (Figure 1-55).

Bottom Plate Outside Corner

Form this corner like the top plate outside corner. Cut one leg of the plate, but fold in or cut off the excess plate to allow the overlapping (Figure 1-56). This type of corner is often used in suspension walls, jigs and soffits.

Figure 1-53. A diagonal brace run from the top plate of a tall wall to just above the spreader and top plate of a shorter intersecting wall. The spreader will be taken off as the tall wall is rocked and the slider is tied in.

Figure 1-54. The first stud of this wall was tied into an already-framed and -rocked wall. Then the top plate of this wall was screwed off to the stud to start the wall. The first stud of a freestanding wall like this can also be shot to structural steel or block walls.

Figure 1–55. Here's the top plate of a freestanding wall fastened in place using a shoe turned up and flattened out.

Figure 1–56. Here the bottom outside corner plate of a soffit is cut and screwed together.

Laying Out Plate

As you lay out the plate to receive the wall studs, it's important to consider the drywall. Place the studs so the drywallers can hang the rock in a series of full sheets without cutting the drywall or adding studs (Figure 1–57). Since most commercial drywall is hung vertically (stood up) with only 4 feet between joints, it's easy to see how much extra work a poor layout job will cause.

To begin the layout, find the point the first sheet of drywall will butt into, and pull your layout from there. The blueprints will call out the spacing of the studs—either 16 or 24 inches on center. If you're pulling the layout off a wall that's not rocked yet, figure the thickness of all the layers of drywall that will cover that wall. Add that thickness to the layout for the first stud. Then clamp the end of your tape measure to this layout mark with a pony clamp and pull the layout from this point, as shown in Figure 1–58. Because the stud layout is commonly marked at either the centers or the soft sides (open sides) of the studs, it's important to indicate which side of the layout mark the studs will sit on. Figure 1–59 shows how to mark the studs for both centers and the soft side. The "X" beside the layout mark tells you the stud sits on that side of the layout mark. Mark the centers with the centerline symbol.

Figure 1–57. An 8-foot sheet of drywall "stood up," breaking perfectly on a structural stud.

Another thing to watch for while pulling layout is to ensure that the drywall doesn't "break" at the edge of door and window jambs. When laying out walls with multiple doors and windows, I recommend pulling an exploratory layout to see where it will hit on the door and window jambs. If the layout hits at unacceptable points around the jambs, burn the necessary number of inches at the beginning of the layout to correct the condition, then mark the first stud layout.

Once you've established the beginning point of the layout, pull layout for the length of your tape measure. Then use a pony clamp to clamp the end of your tape measure to the last layout mark, and lay out the remainder of the wall.

Transfer the layout up for the top plate using the 90-degree reference marks you made while plumbing up the wall line. Your partner will measure from the 90-degree reference mark to the closest layout mark on the bottom plate. Then measure this distance from the 90-degree reference mark at the top plate to establish the first stud layout mark. Because your work area for top plate work is limited to the length of the scaffold, lay out the first stick of plate and stuff the studs simultaneously. Then pull layout from the soft side of the studs for the remainder of the wall.

Figure 1–58. This picture shows the end of a tape measure set to a layout mark and clamped to the bottom plate while pulling stud layout.

Figure 1–59. The bottom plate of a $3^{5}/_{8}$-inch wall is laid out for both centers and soft sides.

Stuffing Studs

Stuffing the stud is a simple, fast-paced step of the framing process. The stuffing process consists of three quick steps. First, insert the stud into the plate diagonally (Figure 1–60), then stand it upright with the stud turned sideways in the plate. Next, slide the stud close to the layout mark, and turn it into place with the soft side (open side) of the stud facing the direction from which the layout was pulled. Finally, tap it into position, exactly on the layout mark, and screw it off.

As you pull the studs out of the bundles and skids (large pallets with several bundles) they'll often be stuck together in pairs. To get them apart quickly and easily, slam the pair of studs on the floor a couple of times. Slamming the studs will jar them apart.

Figure 1–60. As this exterior wall is built, we see a stud being tilted diagonally and stuffed in the top and bottom plate. Then the stud will be stood upright, and set on layout.

When stuffing studs in a soffit or anyplace the bottom plate will be set to a specific elevation later, cut the studs 3/8 inch short, and shove them tight up into the top plate. Soffits and suspension walls are often begun by running the studs wild. They're stuffed in the top plate, then cut to length and plated later. In these situations, it's essential to clamp each stud in place as it's set on layout until they're screwed off.

You've also got to consider cold-rolled channel (CRC) while stuffing studs. CRC is a 16-gauge channel that runs through the stud holes to provide added rigidity or weight support. The stud holes must line up in walls which will receive CRC through the studs, as shown in Figure 1–61. Not all walls require CRC, but always check it out before you start the layout. CRC helps keep the wall straight and resist wind shear. Walls that commonly require CRC are structural stud walls or suspended walls, as well as soffits and wall expansions.

As you stuff walls with CRC, you'll have to lace the CRC as you go. The ends of the CRC need to overlap by 16 inches, as in Figure 1–62. As you stuff the last four to six studs, slide the last stick of CRC loosely into the studs already in place. Once the last studs are stuffed, pull the last stick of CRC through these studs and turn it flat in the stud holes (Figure 1–63). Exterior walls commonly require rows of CRC spaced 4 feet apart, starting at 4 feet above the floor.

In walls where the CRC will carry the weight of the wall via suspension wires, run the CRC through the studs on its edge (Figure 1–64). Under these conditions, the smaller of the two tier stud holes punched in light gauge studs must be up. These holes are designed for the CRC to slide through, to prevent it from sliding around. In suspension work, any movement of the CRC or the suspension wires will affect the elevation of the bottom plate.

Slap Studs and Sliders

Slap studs (or *sliders*, as they're also called) are used in metal stud framing to form an inside corner where two walls intersect. They allow you to build a solid corner using fewer studs by tying the slap stud of one wall to the drywall of

Figure 1–61. Here's a typical 3⁵⁄₈-inch wall framed up, with all the stud holes lined up.

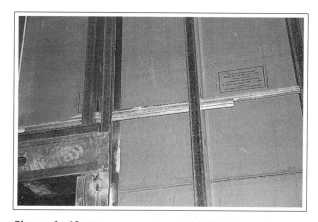

Figure 1–62. You can see the cold-rolled channel run just over a window header (box beam). The CRC joint is overlapped by well over 16 inches, and turned flat in the stud holes.

Figure 1–63. Here the CRC is turned flat in the holes of structural studs and welded to the studs — a common practice.

another. I'll cover the "tying in" process in Chapter 13, *Drywall Methods*.

Regardless of the direction of the layout, the hard side of the slider always faces the wall it will later tie into. In Figure 1–65, the first stud of this partition wall is a slider that will tie into a corridor wall. Notice that the slider saves two studs, compared to the same corner framed with wood studs.

I recommend positioning the hard side of the slider 3/4 inch off of the intersecting wall, and fasten it in place with a single framing screw at the top plate. This prevents the slider from falling out of the wall and becoming damaged beyond use.

Figure 1–64. Notice the 3 5/8-inch studs of a large soffit supported by CRC and suspension wires. The small holes on top lock the CRC in place, helping to maintain the elevation and rigidity of the wall.

Notching Plate and Studs for Obstacles

In some framing situations, you'll have to notch the plate or studs for obstacles such as plumbing lines and the red iron of the superstructure. Cut the notches with your snips on light gauge material. Notch heavy gauge material with a chop saw, a Quickie Saw (a chain saw with a chop saw blade), or torch. Any time you've got to notch the framing material for more than a couple of obstacles, notify your foreman. Ideally (which is rare), the studs should run from plate to plate uninterrupted. If a lot of notching is needed, there's most likely a problem that your foreman may need to write an extra work order for.

Use this process to notch either stud or plate:

Figure 1–65. Here a short wall is tied to a tall wall using a spreader; the first stud of the short wall is the slider. The slider is turned hard side to the tall wall and left unscrewed.

- To begin the notch in light gauge material, mark the outside points of the cut with your snips, as shown in Figure 1–66.
- Next, cut relief cuts approximately 1 inch apart between the marks for the notch (Figure 1–67).
- At one end of the notch, cut into the hard side of the stud to the required depth, forcing the scrap of the cut up and out of the way, as shown in Figure 1–68.
- Turn the corner of the notch by simply forcing the snips to turn while cutting. Continue the cut, cutting lengthwise down the hard side of the stud. The relief cuts made earlier will allow the scrap to simply fold out of the way (Figure 1–69).

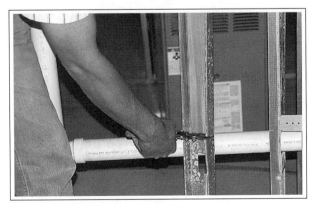

Figure 1–66. Marking the notch with your snips, as shown here, is much faster than measuring and marking the cut with your tape and marker.

Chapter 1: Wall Methods 25

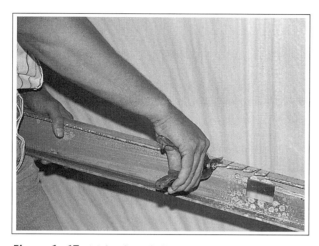

Figure 1–67. Make the relief cuts in the leg of the stud between the two notch cuts.

- As you reach the end of the notch, make the second turn and cut back to the outside of the stud to complete the notch. Figure 1–70 shows a completed notch.

To cut structural studs, follow about the same procedure. Mark the notch with a pencil or marker. Then cut the notch using a chop saw or quickie saw. At the two outside points of the notch, cut the stud to the desired depth (Figure 1–71). With these cuts made, bend the scrap over and out of the way, as shown in Figure 1–72. In Figure 1–73, a quickie saw is used to cut the notch lengthwise down the stud, to completely remove the scrap.

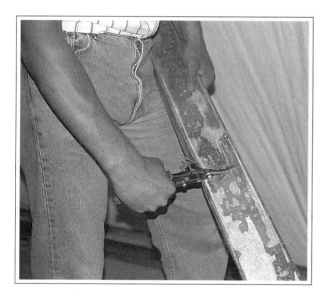

Figure 1–68. Cutting into the hard side of the stud, beginning at the bottom notch mark.

Figure 1–70. The second turn made cutting back out to the edge of the stud to complete the notch.

Figure 1–69. The relief cuts allow the scrap stud to roll up out of the way as the notch is cut out of the stud.

Figure 1–71. Using a Quickie Saw to make the first two cuts of the notch in a structural stud.

Deflection Plate

In this unit we'll discuss two styles of deflection plate, their uses, and their framing methods. The deflection plate is designed to allow the roof or deck of a building to move slightly up or down without damaging the wall.

Deep leg plate, shown in Figure 1-74, is installed just about like the top plate we covered earlier. There are, however, a few significant differences in the techniques used in framing with deep leg plate. First, you'll cut the studs used in the wall 1 to $1^1/_2$ inches short of the deck (Figure 1-75). The second important issue is the way you screw off the studs to the plate. It's acceptable to screw off only one side of the studs to the deflection plate. Whether you're screwing off one or both sides of the studs, the framing screws will be removed as the wall is "topped out." That's essential to allow the plate to lower without damaging the wall. The drywall is also

Figure 1-72. Once the first two cuts of the notch in this bottom plate were made, the notch was beaten over with a hammer.

Figure 1-74. Compare the 6-inch deep leg plate (right) with the standard 6-inch plate (left).

Figure 1-73. Cutting down the hard side of a structural stud using a Quickie Saw.

Figure 1-75. Studs are cut leaving a 1-inch gap in the top plate. The studs are screwed off on only one side. Those screws will be removed as the wall is rocked.

cut 1¹/₂ inches short of the deck, and isn't screwed to the deep leg plate.

The deep leg plate is used in combination with standard plate when called for in the plans. You plate the deck with the deep leg plate, and then slide standard plate up into the deep leg to carry the studs. Fasten the deep leg plate to the deck. Then slide a stick of standard plate the same width as the deep leg, but with 6 inches cut off the length, up into the deep leg plate (Figure 1–76). The 6 inches you cut off the standard plate will stagger the joints of the plate, making it easier to splice the ends together. Slide the standard plate into the deep leg plate, leaving approximately ¹/₂ inch exposed, and clamp the standard plate in place. Now tack the standard plate to the deep leg plate using the appropriate framing screws. You only need two framing screws per side near each end for each stick of standard plate. Splice both types of plate together as you work down the wall.

In many fire walls using this deflection technique, you'll lay safing insulation (dense fire-rated insulation) in between the two types of plate. Cut the safing insulation into strips that are ¹/₄ inch wider than the plate you're using and approximately 1 inch thick. An old filet knife works great to cut the safing. Slide the strips of safing tightly up into the deep leg plate (Figure 1–77). Insulate only as much of the deep leg plate as you can immediately cover by the standard plate. This keeps the dense heavy insulation in place. Once the safing is in place, slide the standard plate in place and work it like regular plate.

Whether you're installing insulation or not, pull the stud layout, marking it to the exposed lip of the standard plate. Cut the studs to allow ¹/₄ to ³/₈ inch play between the studs and standard plate. Then stuff the studs on layout and screw them off on both sides (Figure 1–78). The framing screws connecting the two types of plate will be removed by the rockers as the wall is topped out, and no drywall screws will be run into the deep leg plate.

Accordion plate, shown in Figure 1–79, can be used in any framing that requires deflection. It's specifically designed to give as the deck settles with the added weight of snow and ice. Work the accordion plate following the same procedures

Figure 1–76. This wall is framed to the bar joist using deep leg and standard plate to form a deflection condition. Both sections are spliced, and the lower section is laid out to receive studs.

Figure 1–77. The 3⁵/₈-inch-deep leg plate fastened to the bar joist has the safing installed. The second stick of the deflection plate is being installed to hold the dense, heavy insulation in place.

Figure 1–78. Here the deflection plate is formed using deep leg plate for both sections of top plate. The layout studs are screwed off to only the lower section of top plate.

outlined in this chapter for shooting up the top plate. Cut the studs to within 1/4 to 3/8 inch of the first bend in the plate. Once you've stuffed the studs, screw them off on both sides of the plate, using the appropriate framing screws. The framing screws won't be removed with this type of expansion plate.

Both the *deep leg* and *accordion* plate come in 10-foot lengths and are available in various widths.

Radius Plate

Radius plate is available precut, but you're more likely to cut it from standard plate on the job site, using a chop saw. You can gauge both the depth and spacing of the relief cuts by eye. The relief cuts should come within about 1/8 inch from cutting completely through the hard side of the plate. Space the relief cuts approximately 2 inches apart. The 2-inch spacing allows the plate to conform to nearly all radius wall conditions you'll encounter in metal stud framing.

Use the radius plate alone when plating the bottom of radius walls (Figure 1–80). When plating the deck, cut patterns from plywood or drywall (Figure 1–81). The patterns form sections of the radius, which you'll match at the deck to two plumb points per section of plate. Patterns are also a common method of forming radius soffits and suspension walls. We'll discuss them in detail in Chapter 5.

You form radius plate for arches in much the same way. Perforate the standard plate with relief cuts, but you'll cut both legs of the plate and not the back, as shown in Figure 1–82. Cut the radius plate from either structural or light gauge plate with a chop saw. Again, leave about 2 inches between the relief cuts. Make the cuts square across the plate. If they aren't square, the arch will form a twist in it. In Chapter 6 we'll discuss this process in more depth. Making the relief cuts for either of these methods may require rolling the plate up into the blade. To make the relief cuts, keep the material tight to the fence to keep it square with the blade, and move the plate slowly into the spinning blade to avoid the material kicking back.

Figure 1–79. This is an end view of a stick of 2 1/2-inch accordion plate. The edge of this style of top plate is set to the wall line at the deck just as with any other top plate. Cut the studs to the first bend in the plate and leave the framing screws in as the wall is rocked. The fold in the plate will allow the deflection.

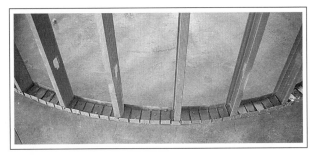

Figure 1–80. A standard 3 5/8-inch plate with relief cuts made approximately every 2 inches is shot to the floor.

Figure 1–81. In this picture, my partner Kelly is clamping the radius plate to the drywall pattern before screwing it off. The pattern will hold the plate in the form of the radius as it's fastened in place to form the top or bottom plate of a drop.

Fastening the Framing Members

One of the big advantages to building with metal studs, for the carpenter, is the speed and ease of connecting the framing members to each other. Often, a couple of quick cuts with your snips and a few framing screws are all it takes. Here we'll look at three commonly-used methods of fastening the materials together.

You can also use this discussion as a reference for other methods that we'll cover in later chapters. As you work to master these cuts, take your time and be patient. The cuts can be tricky, and are often full of sharp edges. Getting in a hurry or getting mad can get you a nasty cut.

Shoes

Cut shoes approximately 3 inches from the end on both stud and plate material. Form the shoe by cutting only the legs of the material and then folding back the end (Figure 1–83). In most cases, you can cut the shoe by eyeballing back the 3 inches from the end of the plate and cutting one leg. But if you're cutting shoes on items like headers, the cuts must be square from side to side. If the cuts aren't square across the hard side of the plate, the shoe will cause the header to twist on the stud it's fastened to. For situations like that, square the cut across the hard side of the plate and cut the other leg. Then fold the end of the plate back to form the shoe.

When you're cutting shoes, more often than not there'll be a shoe at both ends of the material you're using. It's common to give and receive measurements for these cuts as "in-between." For example, the direction "40 inches in between" tells you to cut a shoe, then measure 40 inches and cut another shoe.

Tabs

Tabs are another way you can cut the metal stud material so two pieces can be fastened together (Figure 1–84). Tabs are common on box beam headers, and also to join the intersecting plate of freestanding and suspended wall systems. The length of the tabs will vary, but a 1 1/4-inch tab will do the trick in most situations.

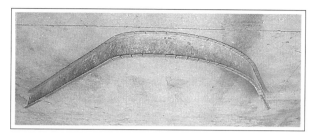

Figure 1–82. You can see a stick of 6-inch plate with relief cuts made in the legs of the plate, allowing it to bend and form an arch. The relief cuts were made on a chop saw to keep the cuts square across the plate.

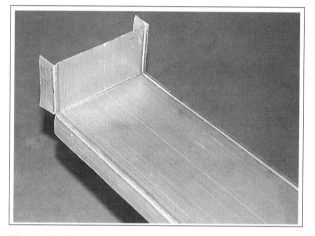

Figure 1–83. A shoe cut on the end of a stick of 25 gauge 6-inch plate. Shoes are used on common headers and kickers.

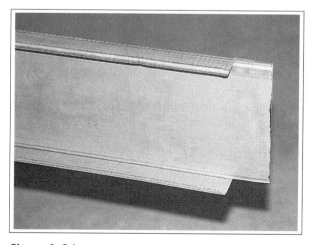

Figure 1–84. Here's a tab cut to the end of a 6-inch stud, which is a common fastening point for spreaders and box beam headers.

Commercial Metal Stud Framing

To cut the tabs, cut through both legs of the material approximately 1¼ inches from the end. Then reposition the material so you can cut down from the end of the material to the first cuts you made in the legs. Cut the tabs with your snips on light gauge material, and use a chop saw on heavy gauge structural material. To make the first cuts 1¼ inches back from the end, set the material in the saw's table and roll it slowly *up into* the spinning chop saw blade. To make the next cuts, reposition either the saw or the material so the saw can cut down the material lengthwise.

Ears

Ears, sometimes used on headers, work the same as a shoe. Cut the ears exactly backwards from a tab: Leave the legs intact and cut away the hard side of the plate (Figure 1–85). To cut the ears, start at the end of the plate and cut down the break (fold) line approximately 1¼ inches. Now, while slowly cutting, turn your snips and cut across the hard side of the plate. Once the cut across the hard side is complete, make a second turn and cut back out to the end of the plate to finish the ear.

Kickers and Other Braces

Here we'll cover three commonly-used braces, some of their uses, and how to cut them. These braces are used for everything from supporting to straightening the framing. Knowing without hesitation which brace to use in a given situation is important. Make sure that the kickers or braces you use are either above ceiling height or inside the framing if you're bracing off of another wall or part of the building.

Kickers

Kickers, like the one shown in Figure 1–86, are useful in countless situations. Use them to brace up freestanding walls, and to straighten as well as strengthen soffits. Cut kickers from stud material, with either one and two shoes cut at the ends. The shoes are often flattened out, with an additional 45-degree bevel cut made to the end of the stud at the base of the shoe (also visible in Figure 1–86). This bevel cut allows the shoe to sit flat against another surface, such as the top plate of a freestanding wall, or the deck.

When figuring the length of a kicker, it's important to consider the angle the kicker will run at. In order to get the maximum strength from a kicker, it must be set as close to a 45-degree angle as possible. The strength of a kicker is also increased by twisting the kicker, as shown in Figure 1–87. Once you've shot the kicker in place, twist it, then fasten the other end to the framing. Another way to increase the strength of the kicker is by screwing the shoe back together at the 3-inch cut points (Figure 1–88). This little spot is hard to get a screw in. The easiest way to fasten it is to set the shoe at the desired angle and clamp it. Then you'll be able to get a screw in it. It's best to do all of this after the kicker's been fastened in place. Which brings up another point: As you fasten the kicker with either screws or powder-actuated pins, place the screws or pins as close as possible to the fold point of the shoe.

Figure 1–85. This shows ears cut on the end of a stick of 6-inch plate. Ears have many uses, but the most common is on common headers on structural framing.

Figure 1–86. A simple kicker cut from 3⅝-inch stud. One shoe has been flattened out and beveled to fit flat to the surface it's fastened to.

Gussets

Gussets are another common style of brace used in the trade. As you can see in Figure 1–89, the gusset is the same as a shoe on the end of a kicker. Gussets are often used to brace furred walls that are built as close as possible to preexisting masonry and precast walls. Cut gussets from scrap stud to whatever size you need. The only side that matters is the one that'll be fastened to the framing. Here's the quickest way to figure it.

Push the end of the scrap stud up against the preexisting wall, then use your snips to cut both legs at about the center of one of the wall studs, or the bottom plate. The other half of the gusset only has to be long enough to fasten to the preexisting wall. Like the shoe on a kicker, screwing the cut points of the shoe on a gusset will make it a lot stronger. If a stronger gusset is required, cut a short piece of stud with ears and screw it off, running it diagonally inside the gusset (Figure 1–90).

Figure 1–87. Here's a 3⁵⁄₈-inch stud used as a kicker. The stud has been twisted before it was clamped to the wall stud to make it stiffer. In contrast, the kicker next to it hasn't been twisted. Both are shot to the bottom of the bar joist.

Figure 1–88. Here you see the cut points on the shoe of a kicker screwed back together to make the kicker stronger above a door jamb.

Figure 1–89. A common gusset cut from 6-inch stud. Rescrewing the legs of the gusset at the cut points will make the brace much stronger.

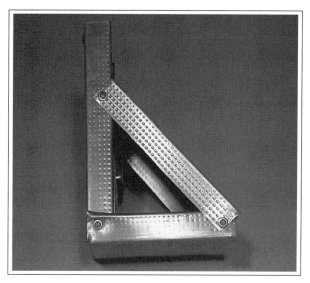

Figure 1–90. Here the gusset itself is braced, making it as strong as a tank. I use this brace on lease (store) walls that will be carry shelves full of heavy merchandise.

The Simple Stud Brace

A simple stud brace is a quick brace that does about the same job as a gusset (Figure 1-91). Cut it from stud material (scrap when possible), then flatten one leg of the stud by hand or with your drywall axe. This allows room to use a shotgun or a screw gun to fasten the brace to the existing wall. The stud brace doesn't provide the strength of a gusset, but it'll make a wall rigid enough for most situations. The biggest drawback of the stud brace is that it's limited to the width of the material it's made from. If there's a 5-inch gap between the new framing and the pre-existing construction, a stud brace made from $3^{5}/_{8}$ inch stud won't work. Large commercial jobs usually have a wide variety of material sizes available, so you won't have much trouble finding what you need. But don't waste too much time looking around. If you can't find what you need quickly, just cut up a stud and move on!

Wall Expansions

Wall expansions are a common part of commercial metal stud framing. They're designed to allow a wall to expand and contract without damaging the wall's finish. At the expansion joint, the wall stops and starts again on the other side of a $3/4$-inch gap in the framing. You'll lay out the expansion joint as you snap the wall lines, and square it across the corridor when working in a hallway. If this work wasn't done in advance, you should know that an expansion joint is required:

- ◆ every 30 linear feet in walls framed from the floor to the deck
- ◆ every 50 feet on ceilings

If you're working in a hallway, use the 3-4-5 squaring method to make sure the expansions are straight across from each other. Expansions that aren't square are hardly noticeable — *until* the walls are completed and the grid ceiling is installed. Then, after it's too late, it stands out like a sore thumb!

To frame the expansion, first establish the center of the expansion (if it wasn't done in advance) and measure $3/8$ inch each way. Plumb these points to the deck, then stop and restart both top and bottom plates at these points. Look at Figure 1-92. The stud layout will also stop and restart at the expansion. Burning 4 inches when reestablishing the stud layout will eliminate a recessed edge of the drywall at the expansion joint. That'll help the trades that follow you.

Set the two expansion studs in the plate, one on each side of the expansion, with the hard side of the studs on

Figure 1-91. A scrap piece of $1^{5}/_{8}$-inch stud used to form a simple stud brace. Notice that the brace is shot off right into the mortar joint.

Figure 1-92. Here's the center of the expansion established at the bottom plate, with the plate stopped $3/8$ inch short of the expansion center. The studs will be set hard side to hard side, with the studs right at the ends of the plate.

the ³/₈-inch expansion layout marks. Then stand up two more studs on the layout. Now cut some pieces of CRC, 5 feet long. Slide the CRC through the stud holes, and stop it when it's through the expansion studs and a layout stud on each side. Lace the CRC through the studs every 4 feet, starting at 4 feet off of the other floor, up to the ceiling height. See Figure 1–93. Get the studs screwed off and the CRC turned flat in the stud holes and you're finished. On fire walls, cut strips of fire-rated safing insulation the width of the framing and stuff it in between the expansion studs.

This wall expansion is a great example of why the stud holes should always line up. Imagine trying to install the CRC if they didn't! Now imagine your job site foreman walking by while you're scratching your head. Keep in mind that even though the expansions are an industry standard, they aren't always framed in. The expansions take extra time to frame in, and once one is done, you're committed. So before you take it upon yourself to correct an obvious oversight, consult your job site foreman.

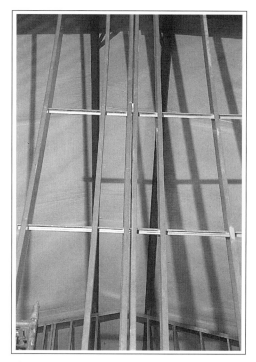

Figure 1–93. With the expansion studs stood up and the layout studs in place, pieces of CRC have been cut to span two layout studs on either side of the expansion. The CRC keeps the wall solid at the expansion.

Straightedging

Straightedging is a commonly-used and widely-acceptable method of transferring a point or wall line quickly and efficiently. A straightedge has many uses, including setting columns and other outside corners as well as finding bumps and dips in walls.

Figure 1–94 shows a drywall rip (the recessed factory edge of a sheet) used as a straightedge to set the outside corner of a column. In Figure 1–95, a level is used as a

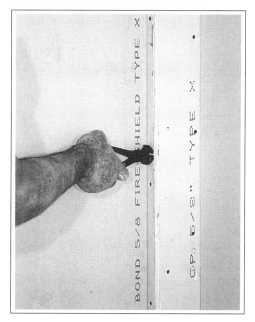

Figure 1–94. Here we see the factory edge of a drywall rip set against the outside corner of a column. Then you can grab a drywall screw left sticking out with a pair of end nippers and use it to pull the column out against the rip. Once set, the opposite side of the column corner is screwed off to lock it in place.

Figure 1–95. Using a level to transfer the wall line of a wall framed to the bar joist on up to the deck.

straightedge to transfer the wall line of a wall framed to the bottom of the bar joist, on up to the metal roofing. When straightedging on framing material, it's important to watch for one of the two ends of the straightedge resting on a framing screw. Of course, this would throw your point off considerably. As shown in these examples, the factory edges of a sheet of drywall, and levels, make good straightedges. A metal stud also makes a good one. These are just a few examples of straightedging. Given a little time on the job, I imagine you'll find uses I haven't even thought of yet!

CHAPTER 2

Headers

Since headers are a necessary part of all metal stud walls, let's learn how to cut them properly. You'll cut the header from plate material, usually forming a shoe on each end. The measurement of a header is commonly given as "in-between." That's the only number you get, since the shoes are always about 3 inches long. If the measurement given is 36 inches, you'll lay out the header at 36 inches plus the 3-inch shoes on each end. When you're framing corridors or other areas with several door or window jambs that are all the same size, lay out a stick of plate with several headers, as shown in Figure 2–1.

It's important to cut the shoes of the header squarely. Until you can eyeball a square cut, use your speed square to square the cut points across the back of the plate. Measure the other shoe of the header from your square mark, or from the shoe after it's cut.

Using your snips, cut the legs of the plate to the square mark you've made across the back of the plate. Then fold the shoe out into position, where it'll be screwed off to the studs. A correctly-cut header with shoes is shown in Figure 2–2. The headers are commonly slipped onto a stud for storage until they're needed.

Not all framers cut headers with shoes on the ends. Some prefer to cut headers with ears on the ends (Figure 2–3). One advantage of cutting ears is that it's not nearly as important for the cuts to be square. We covered the process for cutting ears in Chapter 1. The ears will be about $1^{1}/_{4}$ to $2^{1}/_{2}$ inches long. Cutting the ears longer lets you screw off the header to both the door (or window) studs and the king studs on each side of the rough opening.

You'll find that this style of header is especially useful for structural framing. Structural headers that don't bear any load, such as bottom headers of rough openings, don't need to be box beam headers. You'll have to use a chop saw to cut these structural headers with ears.

However you cut headers for doors and windows, they've got to fit in between the door studs. You can cut headers with ears to fit loosely, allowing up to $1/4$ inch on each side. Headers cut with shoes must be pretty close to exact, and shouldn't have any more than $1/8$-inch play. Measure the length of door and window headers between the door studs right at the jamb.

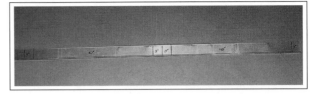

Figure 2–1. You can see a stick of $2^{5}/_{8}$-inch 25-gauge plate that's been laid out with two 40-inch headers, cutting down the time it takes to lay out the headers one at a time.

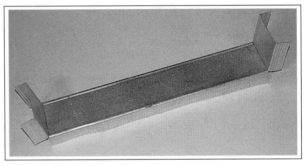

Figure 2–2. A typical header, cut from $2^{5}/_{8}$-inch 25-gauge plate. This is the most common style of header used in nonloadbearing conditions.

Headering Door and Window Jambs

Installing these headers is simple. Just turn them diagonally between the door studs, then turn them back to level as the shoes slide onto the door studs. The process is just the same for windows. As you slide the headers for door and window jambs into place, set a couple of scrap pieces of drywall across the two corners of the jamb to keep the header even. You can see them in Figure 2–4. Screw the header's shoes or ears to the door studs with one framing screw on both sides of each shoe (or ear).

Headering Around Obstacles

As you're building the walls, obstacles like plumbing lines and duct work will interrupt the stud layout. Use headers to overcome the obstacles and keep the stud or studs on layout. Run these headers between the two closest layout studs on

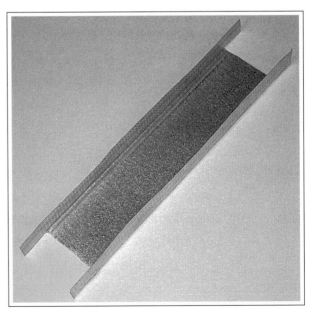

Figure 2–3. In this example the header is cut with ears. This style of header is used in heavy gauge nonloadbearing conditions.

either side of the obstacle, and maintain the layout under and over it (Figure 2–5). Measure the length of the headers between the two closest layout studs, at the top or bottom plate. This will help prevent cutting the header too short or too long, so it bows the studs off layout. Again, the length of the headers will be given as in-between.

After the header is cut to length, you'll slide it in place to within 1/4 inch of most obstacles. However, when headering around duct work, keep the header 1 inch from the duct. This allows room for the collar that the sheet metal worker will add to the duct work later. As you position the header, fasten it in place with the full-length studs pulled tightly into the shoes of the header. A single framing screw on both sides of each shoe will do. Make sure the header is level.

In most situations, you'll frame in a header both above and below the obstacle. It largely depends on whether the wall is a fire, sound or smoke wall and if it'll get topped out. You'll have to check the plans to find out. Even when the wall tops out, many framers opt to skip a top header that carries only one cripple. The drywaller can throw in a scab if they need it. Figure 2–6 shows a properly-headered conduit rack with the cripples in place.

Building Box Beam Headers

The box beam carries the weight of a wall above a rough opening, in both light gauge and structural metal stud framing. It's made up from two studs and two pieces of plate. Cut the studs with tabs on each end. Straight cut the two pieces of plate to a length 1/4 to 3/8 inch shorter than the width of the rough opening you're spanning.

For this example, we'll build a 60-inch box beam out of structural material to span the rough opening of a common window in an exterior wall. Keep in mind that the box beams in exterior walls must be insulated with the same R-rated insulation that will later be used in the wall.

Figure 2–4. Here you can see a typical 3'-0" door jamb with the header slid in place on top of two scrap pieces of drywall. The drywall maintains an even gap between the header and jamb.

Figure 2–5. You can see a large section of ductwork headered in for a 6-inch wall that's been topped out on the opposite side and insulated.

Figure 2–6. Two cripples headered because a rack of conduit prevents the cripples from running to the top plate.

The first thing you'll want to do is get your material cut to length. You need to allow 1/4 to 3/8 inch on the plate, so for a 60-inch rough opening cut the two pieces of plate about 59 3/4 inches long. To cut the tabs on the studs, use the chop saw and make your first cut 1 1/4 inches back from one end. Cut only the legs of the stud. This cut will require that you roll the stud up off the saw table into the blade. Make this cut slowly to prevent the blade from grabbing the stud. From these cuts, measure approximately 59 3/4 inches to the inside cut of the tab on the other end of the stud. Now measure another 1 1/4 inches and cut the stud through. Then turn the partially-cut stud around so it's running parallel with the saw blade. Cut from the end of the stud to your first cuts.

As you build the box beam, you'll need to consider what you're spanning. In this example, we're spanning the rough opening for a standard square window. Only the top plate of the box beam will be turned soft side out (open side of the plate up) to receive cripples. The studs of the box beam will sit in the bottom plate, so the bottom plate will be hard side out. Where both the top and bottom plate of the box beam will receive cripples (in arches, for example), turn out both pieces of plate to receive the cripples.

One of your primary concerns while building the box beam will be keeping the ends and the edges even. To begin, set one of the studs on edge in the bottom piece of plate. Line up the ends of the plate with the inside cuts of the tabs. If these points don't line up evenly from end to end, that's OK — just split the difference. What's important is that the in-between cut points of the two studs are square at the ends. Clamp the two pieces together with clamps at the inside corner of the plate and stud. You can

see that in Figure 2–7. Screw off the stud and plate on both sides of all corners, with the screws spaced approximately 8 to 16 inches apart. Fasten both studs into the bottom plate with the hard side of the studs out. If you're going to insulate the box beam, now's the time. Cut the insulation to the same width as the studs (it's already the same thickness). Cut it to length, then put the insulation into the box beam before setting the top plate in place.

Set the top plate in place on top of the studs with the open side turned out. Clamp it off after making sure the ends and edges are even. Run the screws down through the plate into the edge of the stud, approximately 16 inches on center. Once the plate is screwed off, this box beam's a done deal (Figure 2–8).

Figure 2–7. The two tabbed studs set in the first stick of plate, the ends of the studs lined up with the end of the plate.

Figure 2–8. This box beam's completed and ready to be installed in between the door studs. Only the top plate is turned up to receive cripples. The bottom plate is also turned out when there will be cripples both above and below the box beam.

Stiffbacks

A stiffback is simply a heavy-gauge stud set on edge on top of small metal stud framed ceilings to help carry the weight. The stiffback stud must run perpendicular to the ceiling studs, and span the entire ceiling, as in Figure 2–9. Screw each of the ceiling studs to the stiffback, which will help carry the weight of the ceiling. Just make sure it doesn't interfere with any HVAC or lights. If you cut shoes on the ends of the stiffback and tie them into the wall, it'll make it stronger still.

Figure 2–9. A small ceiling framed with $2^{5}/_{8}$-inch metal studs (called a *hard lid*) is spanned with a $2^{5}/_{8}$-inch stud on top to help carry the weight of the drywall.

Chapter 3

Suspended Drywall Ceiling Methods

Let's take a look at some common "tricks of the trade" for framing a suspended drywall ceiling. I'll also introduce you to some new tools and materials that you should be familiar with. You'll do most ceiling work off of scaffolding or scissors lifts. A scissors lift is a great tool in the hands of a qualified operator. If you've never driven one, have your foreman show you. A scissors lift in unskilled hands can do a lot of damage in a matter of seconds. Never try to snow your foreman. Be straight up with him and he'll respect you for it.

Your main focus in ceiling work will be on maintaining the elevation so you turn out a perfectly flat ceiling. Even ceilings that are on angles must be flat and smooth. Any flaws will cast shadows and make them stand out like a sore thumb.

Another common obstacle to watch for is the HVAC and electrical work, especially the HVAC. As you study the prints, spend some time in the HVAC section. Items on the reflective ceiling plans may not clearly show what's actually up in the ceiling. A simple-looking diffuser on the reflective ceiling print will often break up the ceiling, requiring extra cold rolled channel and suspension wires. Items like this are rarely called out clearly on the prints, but a little detective work early on will save you many headaches later.

Establishing and Maintaining a Constant Elevation

Your first order of business is to establish and maintain a constant elevation that you'll use to build the ceiling (as well as any offsets, angles, or soffits). Here we'll discuss the techniques and equipment you use to establish a benchmark from which you'll establish all elevations. The most commonly-used tools to establish elevations are the laser and the water level. Perhaps you're already familiar with these tools. If not, read on. To do this job, you've got to completely understand these tools and how they're used.

Water Level

The water level is the tool of choice when framing small ceilings. It's quick and easy to set up and use. First, set up the water level reservoir in a secure position, and don't move it again until you've completed the leveling process in the area. When you start, make sure the water level's tubing is capped to prevent leaking. When you remove the cap, the fluid in the tubing will flow through the tubing until it's equal to the elevation of the fluid in the water level's reservoir. When you take off the cap, keep the end of the tubing above the water level canister. If you draw a ring around the tubing with a black magic marker, approximately 6 inches back from the end, it will help you maintain a constant water volume. That makes the water level more accurate. You can add a small amount of food coloring to make the water easier to read, and if you're working where it's very cold, a little rubbing alcohol will keep it from freezing up.

To establish the benchmarks around a room, first get your water level set up and unroll the hose. Take time to check it for air bubbles. If air bubbles appear in the hose, drain the water back into the reservoir, then siphon the water back into the hose. Begin in any corner of the room, holding the hose right in the corners so you can mark both sides of the corner at once. Find the benchmark point by adjusting the volume mark up or down until the water and the volume mark come to an equal point. Then mark both sides of the corner (Figure 3–1). Work your way around the room, marking all inside and outside corners.

With your benchmarks you've established a common elevation around the room. Now you can transfer the ceiling height up from them. First pick a corner, measure up from the floor and mark the ceiling height on the wall, straight up from the benchmarks. Be sure you add the thickness of the drywall to the finish ceiling height on the reflective ceiling print. If you're doing a 9-foot ceiling with $5/8$-inch drywall, mark the frame line at 9 feet $5/8$ inch.

With the ceiling height marked, measure from the benchmark to the frame line mark of the ceiling. To increase the accuracy of your measurements during this process, burn an inch on your tape measure using the 1-inch mark instead of the end of your tape to measure from. Work your way around the room, transferring the ceiling height up from the benchmarks. On walls over 50 feet long, add a benchmark between the corners to prevent the chalk line from sagging in the middle. Once the frame line has been transferred up in all the corners, you're ready to snap it out.

Figure 3-1. While establishing benchmarks around the room, this carpenter marks both sides of the corner right at the volume mark on the water level tube.

Lasers

Lasers do the same job, but they maintain a constant elevation over a much larger area and across the entire plane of the ceiling. Under good conditions you can read the laser beam from 200 feet by eye, and from farther if you use a laser card or beam sensor. You can use a tripod with the laser, but it's usually mounted to a wall bracket which is clamped to a short length of stud screwed to the wall studs in a level position (Figure 3–2). Mount the laser approximately 16 inches below ceiling height, with a safety wire running from the laser to a bat joist. You don't want that $3,000 laser hitting the floor if it falls. You can see that in Figure 3–2.

The head of the laser will self-level, and most models won't light up until the laser has completed the leveling process. The controls for the laser are pretty simple and straightforward. Besides the off/on switch, the laser head rpm and head elevation control for fine elevation adjustments will be about all you need to get going.

To begin, measure up from the floor and mark the ceiling height on the wall close to the laser. Remember to add the thickness of the drywall to the ceiling height. Burn an inch at the ceiling height mark and measure to the center of the laser beam. Work your way around the perimeter of the ceiling, transferring the ceiling frame line up from the center of the laser beam, as shown in Figure 3–3. Again, space the frame line marks approximately 50 feet apart to prevent the chalk line from sagging when snapping out the frame line.

As you work your way around the perimeter of the ceiling, snapping out the frame line as you go, drive a screw into the drywall at the point of the crow's foot of the ceiling frame line. Hook the end of your chalk line to the screw, and reel it out as you roll the scaffold to the next elevation point. Measure out and mark the frame line at this point, then snap a line between the marks. You'll want to pull the chalk line tight here. For a cleaner line, dry snap the chalk line by holding it off the wall and snapping it into the air. Then snap your frame line on the wall. With the frame line snapped, whip your chalk line off the screw you had it hooked to. Now run a screw into the point of the crow's foot where you're working, and roll on.

Figure 3-2. A laser mounted to a wall bracket, then clamped to a stick of utility angle screwed to the wall studs. A length of ceiling suspension wire is used as a safety line in case the laser falls.

Figure 3-3. The framer is "burning an inch" while measuring up from the center of the laser beam.

Computer floors are a complication you'll often find on commercial projects. If your project includes one, you've got to take it into consideration when establishing the ceiling height using either of these methods. A computer floor is a raised flooring system that's usually about 2 feet up off the slab. The elevation of the computer floor is established by the installer or the general contractor, and you'll measure all ceiling elevations from the computer floor height.

Utility Angle

You'll use utility (or wall) angle to form the perimeter of the ceiling, with the angle sitting on your chalk line and following it exactly. Screw the utility angle (Figure 3–4) to the wall studs through the drywall, using drywall screws. Drive the screws in slowly with some control, to avoid stripping them out. *These screws must not strip out.* The utility angle is all that supports the ceiling for the first 4 feet, so it's very important for it to be securely fastened to the wall.

The joints are also very important and must be overlapped. This is best done by overlapping the end of the piece you're setting in place over the last wall stud of the stick already screwed to the wall, as shown in Figure 3–5. Then screw the two pieces together with a few sharp-point framing screws before you screw them off to the wall stud. Whenever you screw through two pieces of framing material and then to a metal stud material like this, they'll try to kick apart. Overcome this problem by predrilling a pilot hole with an S-12. That's a self-drilling framing screw (also called a *lath screw* or *34*). Or you can drive a drywall screw through the utility angle with your screw gun in reverse. Then use a new screw to screw it to the stud.

To form inside and outside corners with the utility angle, cut one leg of the angle at the corner point, allowing the angle to fold over and form the corner. It's also common to simply cut the angle off at the corner, and screw it to the corner stud. Both of the methods are equally useful, and you'll come across many opportunities to use them both.

Figure 3-4. You can see a stick of utility angle screwed to the studs of a tapped-out and fire-taped wall, with several sticks of DWC screwed to it.

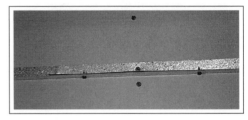

Figure 3-5. Here the utility angle has been overlapped over a wall stud, allowing both sticks of utility angle to be screwed to the stud.

Tying Suspension Wires to Bar Joist

As we start this discussion, let me introduce what may be a new tool to you. You'll use a *wire tier* to tie the suspension wires. You can see the wire tier in Figure 3–6, where it's fastened to the end of a length of conduit with an S-12. You can usually get an electrician to bend a crank handle on one end of the conduit, making it quick and easy to use. The suspension (or hanger) wires (which are usually #9 wire) are bent over approximately 8 inches from one end. You'll use this bend to tie the wire to the tops of the bar joist, and also to store the wires while working, hanging the wires on the handrail of your scaffold or scissors lift (Figure 3–7).

Tie the suspension wires over the tops of the bar joist, after you've guided them into place with your wire tier. First, hook the bend of the wire loosely to your wire tier, and then guide the wire into place over the top of the bar

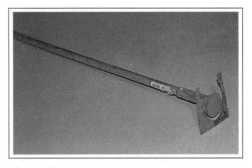

Figure 3-6. A wire tier that's fastened to the end of a length of conduit with a self-drilling framing screw.

Figure 3-7. Several suspension wires have been bent and hung from the handrail of a scissors lift.

joist (Figure 3-8). Once the wire is in position, push it sharply over the top of the bar joist, through the gap between the bar joist and the corrugated metal decking. When you've shoved the bend of the wire all the way through the gap, pull back on the wire lightly until the bend catches on the bar joist. Now slide both sides of the wire into the wide opening of the wire tier, then use the crank handle to spin the wire tier and tie the wire (Figure 3-9). Wind the wires four or five times to make sure it won't come loose as it's being stretched during later steps of the work.

If you can get up to the bar joist, you can also tie the suspension wires by hand. When tying the wires to the bar joist, be sure they're tied tight with at least four wraps. Do yourself a favor and wear some gloves when tying wires by hand.

Here's another common technique for tying up the suspension wire. Use a *deck punch* (large pliers punch) or a common carpenter's punch to stab a hole on each side of the corrugated metal deck. This method will work whenever you're tying the wires to the roof decking. Punch the holes in the decking and slip the end of the wire through the holes, then tie it (Figure 3-10).

Space the suspension wires in neat rows 4 feet apart, starting 4 feet off of a perimeter wall. The bar joists are usually spaced 4 feet apart. If you use the cross bracing in the trusses as a guide for the wire placement, it'll help you keep the rows even and straight. This 4- x 4-foot pattern for the suspension wires is the standard spacing for nearly all suspended ceiling systems.

You'll use a much more complex wire pattern and framing system when framing a *seismic ceiling*. This complex suspended drywall ceiling system is common in regions of the country where there's frequent earthquake activity—like California. But for now, we'll focus on the basic ceiling.

Splay Wire

Use a splay wire when there's no place to tie a suspension wire directly above where it's needed. Like a typical suspension wire, tie the splay wire to the top of the bar joist, closest to the point where it's needed. Then the wire will run at an angle to the point where it will be saddle-tied to the cold-rolled channel (or CRC). See Figure 3-11. Running the splay wire at an angle often means the wire to be too short to tie a proper saddle knot. If the end of the splay wire doesn't run (on its angle) at least 12 inches past the CRC, it must be spliced. Splicing the suspension wires is the topic of our next discussion, so let's move on.

Figure 3-8. The suspension wire guided up to the top of the bar joist, hooked on the wire tier. Once in position, give the wire a sharp shove to get it up over the top of the bar joist.

Figure 3-9. Here the wire has been shoved over the top of the bar joist, then both sides of the wire grabbed in the large opening. With the wire in the jaws of the wire tier, the wire tier is spun to wind up the wire.

Figure 3-10. With the holes punched in both sides of the corrugated decking, the wire is poked through and tied as normal and then stretched.

Commercial Metal Stud Framing

Figure 3-11. On the right you can see a suspension wire hanging straight down from the bar joist, while on the left the splay wire runs at an angle. If you look closely at the splay wire where it crosses the sprinkler pipe, you'll see that it's spliced.

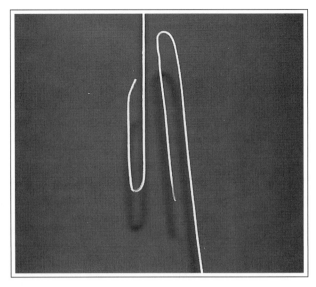

Figure 3-12. The first bends made to the ends of the suspension wires. The wire tied to the bar joist is bent about 4 inches from its end, and the wire being spliced into it (bent down) is bent about 6 inches from its end.

Splicing Suspension Wires

Splicing the suspension wires is the standard technique for dealing with wires that are too short to properly saddle tie to the cold rolled channel (also called *black iron* by the old timers). You'll encounter many situations that will require splicing a short wire, especially when running a splay wire. When you're splicing a wire, add a full strand, with the splice at least 16 inches above the CRC. Then you can cut it shorter after the splice is complete if needed. This prevents the splice itself from getting in the way later when you saddle-tie the wire.

To splice the two wires together, first bend the ends of the wires over sharply. Bend the wire tied to the bar joist approximately 4 inches back from the end. Then bend the wire being spliced to it about 6 inches back from its end. To complete the bending process, put about a 1-inch bend on the tip of each wire perpendicular to the first bends you made. You can see it in Figure 3–12. The bends at the tips of the wires make it easier to tie the splice, allowing the wires to tuck under each other. And that's what we're ready to do. As we begin this step, I'm going to designate the wires *A* and *B*, with *A* being the top wire tied to the bar joist. Keep in mind that we're simply tying a square knot with the wires.

As you read these instructions, follow along on Figure 3–13, going from left to right. To begin, put the wires together back to back, with

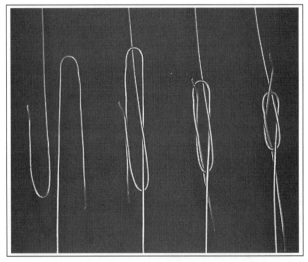

Figure 3-13. From left to right: First, the bends of the suspension wires set back to back, ready to start the splice. Then the wires wrapped around each other. Next, the tip of the A wire pushed through the bend of the B wire, forming the knot. Finally, the completed splice, with the spliced wire ready to be stretched and saddle-tied to the CRC.

the tip of wire *B* just above the bend in wire *A*. With the wires together, spin the longer wire *B* 360 degrees, wrapping around wire *A*. Finally, to complete the knot, push the tip of wire *A* through the bend in wire *B* as you pull down on the bottom wire, cinching up the knot. You can see the completed knot at the right in Figure 3–13. Pull the knot as tight as you can by hand for now. The knot will get cinched down tight as the wires are stretched later.

Bridging the Bar Joist

Another common way to get a wire where you need it is to bridge the bar joist, which allows the wire to hang straight down to the point where it's needed. What we'll basically do here is make a miniature I-beam out of two short pieces of CRC and set it on top of the bar joist with a suspension wire tied to it.

First, determine the length of your CRC by measuring from center to center of the bar joist that you need to span. Then cut the two pieces of CRC to length and screw them together back to back with S-12s, spaced approximately 12 inches apart. Now tie the suspension wire to the CRC, just like you tied it to the top of the bar joist.

Your next move is to slide the CRC beam into place on top of the bar joist. Set it on its edge, as shown in Figure 3–14. This makes for a very tight fit between the bar joist and the metal decking. To get the CRC in place, slide it up into one bar joist just enough to clear the lip of the other bar joist. Then slide the CRC up onto this side until the ends are roughly in the centers of the two bar joists.

Suspension Trapeze

A suspension trapeze is yet another way to get a suspension wire where you need it. This method is typically used to overcome obstructions such as large HVAC trunk lines and wide racks of electrical conduit. The trapeze is simply two suspension wires tied to the bar joist on both sides of the obstruction, and then stretched. Next, cut a piece of CRC long enough to span the obstacle and the two suspension wires, and tie a suspension wire to the CRC where it's needed. If the CRC looks like it might bow, double it up as we did in our last discussion. In either case, tie the CRC to the suspension wires using the saddle knot covered later. Level the CRC by eye 2 or 3 inches below the obstacle you're spanning to give you some room to work. Next, tie the suspension wire to the trapeze where it's needed, just as you'd tie it to the bar joist. Figure 3–15 shows a completed trapeze. The suspension wire left hanging will be tied later to the CRC of the suspended ceiling.

Stretching and Bending Suspension Wires

Stretching the suspension (or hanger) wires properly is a very important step in suspension work. Wires that aren't properly stretched at this point of the job may stretch later under the weight of the ceiling, allowing the work to sag. We'll also take a look at bending the hanger wires, which in ceiling work is a common step in

Figure 3-14. The two pieces of CRC are screwed together back to back with S-12s. The wire was tied to the CRC before it was slid up over the tops of the bar joist. Once in place, the suspension wire is stretched and saddle-tied just like any other suspension wire.

Figure 3-15. Two pieces of CRC, screwed together and supported by a suspension wire on each end. Then a suspension wire is tied to the trapeze to support the ceiling.

the stretching process. Keep in mind that *all suspension wires must be stretched.*

Stretch the wires by lightly biting into the wire with the wire cutters on your lineman's pliers, about 20 inches below the ceiling height. Then use your hammer or drywall axe to hammer down with firm blows on the hinge of your lineman's pliers, as shown in Figure 3–16. To stretch the wire properly, hit down on your pliers three or four times, and then cut the wire off where you bit into it with your pliers. Wires that have been spliced will need a few extra strokes to cinch down the knot and stretch both wires.

Bending the wires as they're stretched is common in ceiling work. These bends are used to temporarily carry the CRC as it's hung. Usually, every other wire is bent, depending on the length of the CRC. Bend two wires for each stick of CRC. Use the laser to maintain the bend point of the wires. Your first move is to determine the bend point of the wires (which is the bottom of the CRC). To figure the bend point, measure from the center of the laser beam to the utility angle. We'll call it 10 inches in this example. Then add the thickness of the drywall channel ($7/8$ inch) minus $1/4$ inch for adjustments:

$$10" + 7/8" - 1/4" = 11 1/8"$$

The accuracy of the measurement can (as always) be improved by burning an inch. When we do that, the final number is $11 5/8$ inches from the center of the laser beam to the bend point of the wire.

Now let's put our sample measurement to work. Pull your tape out and lock it about 6 inches past the measurement. For this example, lock it in at 18 inches. Hold your tape measure by the tape itself, at about the 14-inch mark, with your thumb and index finger. You can grab the wire with your free fingers, as shown in Figure 3-17. Slide your tape up or down the wire until the center of the laser beam hits your number—$12 1/8$ inches, in this case. Keep your wire pulled tight, and somewhere in the vicinity of plumb.

Now use your lineman's pliers and grab the wire at the 1-inch mark on your tape, also shown in Figure 3–17. Drop your tape into your nail pouch and bend the wire

Figure 3-16. Gripping the suspension wire tightly with the cutters of the lineman's pliers while hammering down on the pliers to stretch the wire.

Figure 3-17. Burning an inch while measuring up from the laser beam to the bend point (split point) on the wire.

sharply into a horizontal position, right at your pliers. Next, drop your pliers in your pouch and grab the wire just above the bend with your right hand so that your little finger is right at the bend. With your left hand, grab the wire and fold it up right around your little finger, forming about a 15-degree angle. Slide your left hand up the bend approximately 10 inches and bend the wire back down, forming about a 30-degree angle. From this last bend, eyeball about 10 inches and cut off the wire (Figure 3–18). You can make these last two bends simultaneously after you get the feel for it.

Saddle-Tie Suspension Wires to CRC

Saddle-tying the suspension wires to the CRC is the usual way to support the CRC, which in turn will carry the entire suspension system. Some of the wires you'll tie will have been bent in advance as the wires were stretched. Use these bent wires to temporarily carry the CRC as you're tying the wires. The wires are saddle-tied the same way, whether they've been prebent or not. The bent wires just give you something to grab on to while you're tying the wires. In some cases, the wires will already be stretched when you get to this step, but if you have any doubt, stretch the wires before you tie them.

Whether the wire is bent in advance or straight, the tying process will begin with the wire behind the CRC. Unbent wires should hang no more than 16 inches below the CRC. For this example, the unbent wire was cut to length as it was stretched. In ceiling work, you tie the bent wires first, then tie the straight wires before you move on. As we walk through the saddle-tying process, we'll tie an unbent wire to show the entire process. The only difference is that the prebent wires have the first bend already made.

Your first move is to bend the wire into a horizontal position across the bottom edge of the CRC, as shown in Figure 3–19. Next, bend the wire back up the front side of the CRC, over the top and sharply around the back of the wire. Check Figure 3–20 to make sure you're on cue. Now

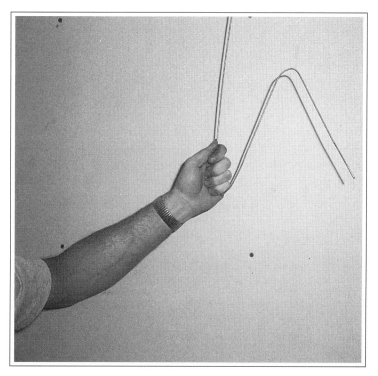

Figure 3-18. Here's the suspension wire, stretched and prebent.

Figure 3-19. With the bend point (elevation) of the suspension wire determined and cut to length, the wire has been bent into a horizontal position.

fold the wire back down the front of the CRC, around the bottom and up the back side. Then wrap the wire around itself at least three times, working up, as in Figure 3-21. In cases where the ceiling will support a heavy weight, you'll have to double saddle-tie the wires. Just let the wire hang a little farther past the CRC and follow this pattern around twice.

Fastening Ceiling Materials

Here we'll take a look at some of the common methods of fastening the ceiling materials together, using either tie wire or framing screws. The prints will usually call out whether to use screws or tie wire. If they don't, your foreman will make the call. It'll usually be tie wire, because it's cheaper than screws. Using tie wire also eliminates the need to drag a cord and screw gun around with you. Once you've mastered the tying technique, you'll find that you can tie as fast as you can screw the materials together.

As we cover the procedure for tying the ceiling materials together, there are a couple of key factors to keep in mind. First, when you tie the drywall channel to the CRC, use two wires at each tie point. While these wires must be tight, you have to be careful not to break them. It's a fine line, using enough pressure to tighten them without breaking them. Second, it's important that the winding of the wires be off to the side of the drywall channel. If it's not, the windings will deform the drywall as it's hung later.

That brings us to a new tool. *End nippers* are used to wind, cinch down, and cut the tie wires. A good pair of end nippers will cost you about 20 bucks. Let me once again advocate American-made tools. After 15 years in the trades, experience has taught me they're far superior.

Tie Wire Method

To tie the drywall channel (DWC) to the CRC, first you need to bend the 18 ga. tie wire. Take a small wad of tie wires about the size of a dime and bend them over approximately 6 inches from one end. Keep in mind that you'll make all of the bends with the *long* end of the wires, keeping the short end stationary. The bends in the wires are an important part of the tying procedure. As an added bonus, they also make the wires easy to store while working because you can hang them from the hammer loop on your tool belt. Remember, you'll use two wires at each tie point, so load up 20 or so wires at a time to reduce the number of times you need to stop

Figure 3-20. The suspension wire bent tightly up the face of the CRC, then sharply around the back of the wire.

Figure 3-21. Here the suspension wire is wrapped back down the face and up the back of the CRC, then wound around itself four times.

working and reload your belt. Store additional wires on the handrail of your scaffold.

Begin the tying procedure by facing the side of the DWC, looking parallel down the CRC. First hook the bend of the two wires over the CRC on the far side of the DWC as you hold the DWC tightly to the CRC on layout with your free hand (Figure 3–22). With your working hand, bend the long side of the wire catercorner across the face of the DWC, slipping the wire under the thumb of your free hand to hold it in place. See Figure 3–23. Next, bend the tie wires up over the CRC and back across the face of the DWC catercorner to the short end of the wire, then wind them up (Figure 3–24).

Once the wires have been wound, use your end nippers to cinch down the wires. To do this, lightly bite into the wires at the bottom of the wrap and wind the slack out of the wires. Now pry more slack out of the wires by bracing the face of the end nippers against the DWC and prying (Figure 3–25). Then you wind up the slack

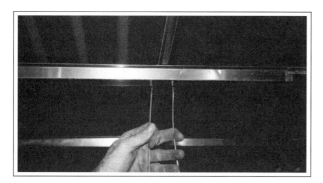

Figure 3-22. Beginning the tie by hanging the tie wires over the CRC on the far side of the DWC.

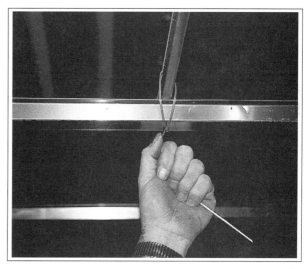

Figure 3-24. The tie wire bent over the top of the CRC and catercorner back across the face of the DWC to the short end of the wires.

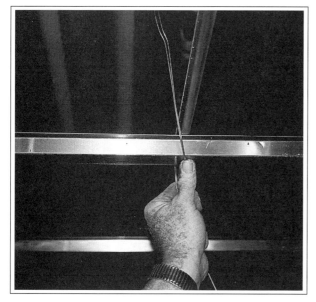

Figure 3-23. Bending the wire catercorner across the face of the DWC. Notice the thumb holding the wire tight to the DWC, preparing to reach over the CRC to bend the wire over the CRC and back down.

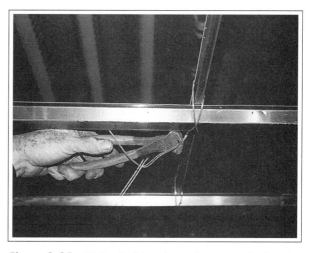

Figure 3-25. While cinching down the wires, the head of the end nippers is wedged against the DWC to pry the slack out of the wires. The slack in the wires is then wound up tight.

until the wire is tight again. Repeat the prying and winding process one or two times, more if needed, to get the DWC cinched tight to the CRC. Just be careful not to break the wires. To check the tightness of the materials, use your axe or hammer handle and tap the DWC next to the wire. A rattle in the material will tell you the wire needs to be cinched down and wound again.

When it's tight and the tie complete, bend the knot off to the side of the DWC and cut off the excess wire. Then tap the DWC over to the layout mark on the CRC if needed. Figure 3-26 shows the DWC properly tied to the CRC, and on layout. If just one of the wires breaks while you're cinching it up (and they will), start over with two new wires.

Screw Method

The second method is to screw the DWC (also referred to as *hat channel*) to the CRC using S-12s. This is also a common way to fasten the materials together. In this situation you'll simply hold the DWC against the CRC on layout, and screw both sides of the DWC in place. See Figure 3-27. When screwing the DWC off, wear safety glasses to protect your eyes from the hot burrs that fly off the material. Cupping your free hand under the screw as you're running it into place helps prevent these hot burrs from going down your shirt (ouch!)

Joints in CRC and DWC

Fasten joints in the CRC and DWC using tie wire. In both situations, one tie wire folded in half will do the job. Double the tie wire over and tightly wrap it twice around the joint. Then wind it on the side of the material (Figure 3-28). Using your end nippers again, wind the slack out of the wire, then pry some more slack in the wire and wind it back out, until the joint is tight. Check the joints in the materials by tapping them with your axe handle to ensure they're tight. Tie the joints in either CRC or DWC at both ends of the joint or every 12 inches on long overlaps.

You can also fasten the joints in the CRC and DWC by screwing them together with framing screws. The CRC will require S-12s fastened at both ends and the middle on both the top and bottom of the joint. Screw the light gauge DWC together at both ends of the joints with sharp-point framing screws, on both sides of the material.

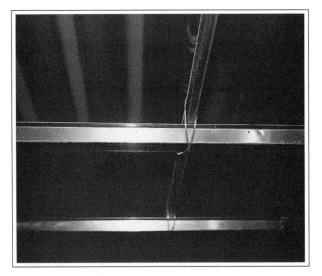

Figure 3-26. The completed tie.

Figure 3-27. The DWC screwed to the CRC using the self-drilling framing screws.

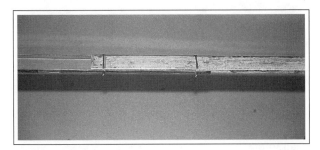

Figure 3-28. A joint in the CRC properly tied with tie wire.

Corners in CRC

Because the CRC is the chief source of rigid support in a suspension system, it's important that it be tied together properly at joints, including the corners. That's what we'll look at here. Basically, you'll cut a shoe on an end of a stick of CRC, and then either screw the two pieces together with S-12s, or tie them together with tie wire. To make the needed cuts on the CRC, you *should* use a pair of bullnose cutters — but your snips will work. Many framers keep a worn-out pair of tin snips in their bucket, just for cutting heavy gauge material.

Forming the corner itself will require a bent, squarely-cut shoe on one stick of CRC that allows it to butt together either back to back or open side to open side. In either case, tie or screw the CRC together. However, when you form a corner with the material back to back, it must be screwed off. If you use tie wire, the edges of the CRC will cut it, weakening the joint. Cut the corner (or shoe) approximately 12 inches long, and screw the two outside points and the middle. It's important to keep the bottom edge of the two materials even. Use a clamp if needed, but don't get in the habit of clamping materials together unnecessarily; it's a waste of time.

Forming corners with the CRC hitting open side to open side is much more common, and requires just a little forethought as you hang the intersecting sticks of CRC in place. This style of corner is also a stronger corner, as the lip of each piece of CRC shares the weight load. Again, you'll cut the shoe about 12 inches long. Just eyeball your cut. It doesn't have to be exact. Tie the cold-rolled corner with the tie wire, following the same procedure outlined earlier. This type of corner is also screwed together using S-12s, with screws at both ends of the shoe on both the top and bottom of the CRC. You can see them in Figure 3-29.

Additional Support for Suspended Ceilings

In many suspended ceilings, regardless of size and shape, you'll have to provide additional support. That means additional CRC and suspension wires. Often, a large flat ceiling is only the base for an elaborate ceiling, with all the additional work framed to and supported by the flat ceiling.

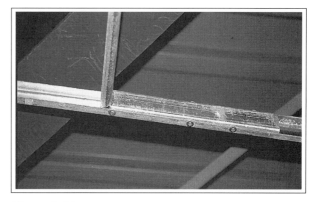

Figure 3-29. Here you see the shoe cut on the end of one stick of CRC set inside the lip of the intersecting row of CRC and screwed off.

You may also have large HVAC diffusers and vents that penetrate your ceiling and require additional support on both sides of the penetration. Recognizing these conditions before you start the work will save you lots of time down the road. As you study the reflective ceiling section in the prints, take time to compare it to the HVAC and electrical plans.

When you do have to provide extra support, it's best to tie your extra suspensions wires to the bar joist above where they'll be needed when you're hanging the rest of the wires. But if a specialty condition slips past you when you're planning the job, a full-length stick of conduit for your wire tier should get your wires up to the bar joist. Add the extra CRC after the drywall channel work has been completed, and you've laid out the vent opening or framing line on the DWC. Then set the extra CRC exactly where it's needed, and tie or screw it in place. Now stretch and saddle-tie the extra suspension wires to the CRC. When it's tied, set the wires to the laser and cut the vent opening out in the DWC with a reciprocating saw. Figure 3-30 shows a typical situation where additional CRC and suspension wires were added.

Tuning in the Ceiling Elevation

Tuning in (or dialing in) a suspended ceiling is one of the final steps of framing the suspended ceiling system. For this step you'll need the laser

Figure 3-30. Here the extra CRC is running the same direction as the bar joist in order to miss the lights and duct openings.

Figure 3-31. The suspension wire has been saddle-tied to the CRC and locked in place using a self-drilling framing screw.

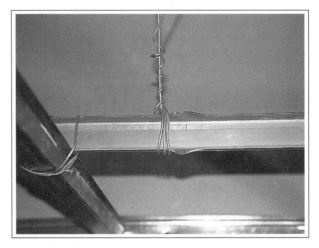

Figure 3-32. The suspension wire tied to the CRC with a saddle and a half knot (wrapping the wire around the CRC one more time). The legs of the CRC have been spread apart, locking the suspension wire in place at the desired elevation.

again, as you'll set the ceiling to the laser beam. The measurement between the drywall channel and the center of the laser beam will be the same as between the utility angle and the laser beam. If everything works out right, all needed adjustments will be to raise the ceiling.

To raise the ceiling, slide the suspension wires on the CRC from their original position to out of plumb. The farther out of plumb you slide the wire, the more the ceiling will rise, until you get the ceiling to the predetermined elevation. When you check the elevation of the ceiling while tuning it in, pull down firmly on the CRC closest to the drywall channel you're measuring from, imitating the weight of drywall that will be hung on the ceiling later. Remember, as you check the ceiling elevation, you'll measure from the drywall channel to the center of the laser beam.

Once you've achieved the desired elevation, lock the wire in place. There are a couple of ways to do that. First, you can run an S-12 partially into the bottom of the CRC, right next to the wire (Figure 3–31). Or you can spread out the legs of the CRC with your lineman's pliers (Figure 3-32). The hump that creates in the CRC will prevent the wire from sliding back. Bending the CRC is the quicker and more common of the two methods used to lock the wires in position.

Chapter 4

Soffit Methods

In most cases, the layout work will be done in advance by your foreman or the general contractor's layout man. But when it's not, you'll have to be prepared to lay out the soffit. One of the most important things you need to know is where to find the soffit dimensions in the prints. There are four sections in the prints that show the soffit dimensions: the reflective ceiling section, the casework section, and the floor plan will cover the interior soffits. The elevation section will show the outside soffits and canopy work. Each of these sections will call out more detailed drawings for any specialty items: stair-step soffits, jigs and light troughs, for example.

Laying Out a Soffit

The majority of soffits in smaller rooms, like offices, are only there to fill the void above cabinets (casework). These soffits, or *drops* as they're also known, bridge the recessed portion of a wall that allows a cabinet to sit back inside a wall. In this situation, the only number you'll need is the soffit's elevation. Then you'll simply match the vertical face of the soffit to the frame line of the wall studs already framed (Figure 4–1). Even if the wall is out of plumb, the soffit will follow it. This is a good example of being married to a pre-existing condition. Your chief concern is that the soffit face blends smoothly into the wall.

For other small soffits, you'll simply lay them out off of the finish walls in the room. You won't do the soffit work until the walls have been rocked and taped to allow access, especially when the walls top out (rock to the deck) and are fire taped.

Pull the layout for a large soffit following the same procedures you use to lay out walls. First establish the soffit frame line off of a wall line, grid line or control line (*gospel line*) to make it jibe with the surrounding work. You can also use column centers and window mullions to establish the soffit layout. Large soffits are often the centerpiece of a large area. In most cases, the layout work will be done in advance.

Soffit Elevation

There are several ways to establish the elevation of a soffit. In smaller rooms and offices, a water level (or even your 4-foot level) will do the job. As you establish the elevation, remember to add the thickness of the drywall in order to get your frame line. Larger areas will nearly always have an elevation established by the general contractor, since they do the flat work (concrete). When possible, make them establish the elevation. That way, the general contractor (instead of your outfit) is liable for any problems concerning elevations.

You'll find the benchmarks on the red iron piers (columns) of the superstructure. Then use a laser to figure your soffit elevation from those benchmarks. This will require a little math. Measure from the benchmark to the center of the laser beam. For this example, let's call it a difference of 5 feet. Add to this the benchmark elevation. If the benchmark is at 4 feet, the total elevation would be 9 feet. Next measure, from the soffit (or any) elevation, which we'll call 10 feet, for a difference of 1 foot. To your number, add the thickness of the drywall, let's say $5/8$ inch, so you have 1 foot $5/8$ inch from the center of your laser beam to the bottom of the soffit frame line.

Jigs

One of the great advantages of framing with jigs is that no matter how intricate the soffit, you'll do 90 percent of the layout work on the jig table. Your chief layout concerns will be the elevation and a single plate line that will be either the inside (front) or the outside (back) of the soffit. Lay out the soffit plate line on the floor, pulling it off of a wall line or control line, and then plumb it up. You can also pull the soffit frame line off of other suspension work already framed, like a suspension wall or a wall framed to the deck.

Figure 4–1. This is the end stud of the soffit tied into an already-framed and- rocked wall. The soffit end stud is simply lined up with the back side of the drywall on the wall.

Figuring Soffit Stud Length

There are several different ways to determine your stud length, depending on the framing conditions you run into. Of course, using a laser on the job will take the hassle out of this and all the rest of your elevation work. But framing a soffit to a suspended ceiling is just as good as having a laser, because you'll have a constant elevation. Then all you need to do is figure the difference between the ceiling height and the bottom elevation of the soffit, and allow $3/8$ inch for adjustments. There's your stud length. Now let's take a look at a couple of other methods.

Framing to the Deck

When framing a soffit to the deck, there's no telling what conditions you'll run into. While the deck (or *hardlid*) is occasionally flat and level, it's usually not. It'll often have offsets of several inches. See Figure 4–2. Once the top plate is fastened in place, use the benchmarks made at both ends of the soffit to maintain the desired elevation. First, measure from the floor, allowing for the drywall. Mark the frame line elevation at one end of the soffit. Now, at the same end, measure from the benchmark to the soffit frame line elevation. Then use this number at the opposite end of the soffit, measuring up from the benchmark to the frame line elevation (bottom plate line) of the soffit.

When you've established the frame line elevation at each end of the soffit, measure from the frame line to the deck at each end to see how level the deck is. If the deck is reasonably level, you're ready to go. Just measure from the deck down to your plate line (frame line), and subtract $3/8$ inch for the bottom plate adjustment to find the stud length. If the deck is only slightly out of level ($1/2$ inch or so) use the shortest number and subtract $1/8$ inch instead of $3/8$ inch.

Running Studs Wild

When framing long soffits and suspension walls with long vertical studs, it's common to use full-length studs and run the studs wild, then cut them to length later. Select a stud length that will hang from the top plate to below the bottom plate elevation, with no more waste than necessary. When the top plate is laid out, clamp

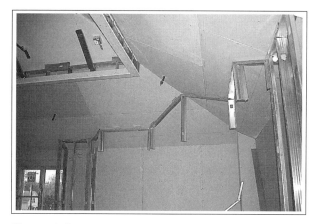

Figure 4–2. This drop is full of angles, and it had to be framed to a vaulted ceiling with the center of the drop right at the center (hip) of the ceiling. I set up a laser and used that to maintain the stud lengths on this drop.

three or four studs at a time on layout, and screw them off to both the front and back of the plate. Stud up the entire top plate, taking care that the stud holes line up, and install the cold-rolled channel as needed. You'll often need a horizontal control stud to keep the studs straight during the framing process. This method is covered in detail later in this chapter.

Set up the laser and establish your plate line, measuring up from the floor or benchmark. Add an additional $3/8$ inch for the bottom plate adjustment and measure from this point down to the center of the laser beam. Now transfer this number from the center of the laser beam to each stud in the wall. As you mark the cut point on each stud, square it across the back of the stud with your speed square and "score" the stud with your razor knife. Then use your snips to cut the legs of the stud to the score line. Now fold the bottom of the stud back and forth three or four times until it snaps off along the line you scored earlier. See Figure 4–3.

In each of these examples, there's a common process. Establish the plate line elevation, and allow $3/8$ inch for the bottom plate adjustment. Then figure the difference between this point and the top plate to find the stud length. There are too many variables in the trade to cover all the possible conditions you'll encounter, but the examples covered here are adaptable to nearly any situation you'll run into. I can't emphasize enough the value of the laser. In fact, you're

Figure 4-3. You can see two studs squarely cut to the marks, and the cut line marked on the three uncut studs. They're all held in place with a horizontal control stud.

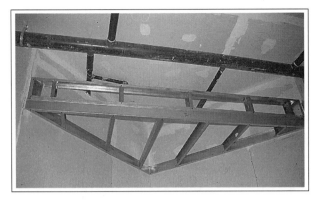

Figure 4-4. This small drop was framed just high enough to get above ceiling height.

sure to run into situations where a laser is the only way to get the job done.

Your Call

In many casework soffits, such as the one in Figure 4-4, the soffit isn't framed all the way to the deck. These soffits commonly run only from the bottom finish elevation to 6 or 8 inches above the ceiling height. You'll simply establish your elevations and measure for your stud length, which only has to get up over the ceiling height.

Plumbing the Corner Studs

In many soffit framing situations, the soffit will tie into walls that have already been framed and rocked. You'll simply tie the end studs into the drywall (or a wall stud behind the drywall when possible), which will hold the end studs plumb. The plumb work here is all done with your level. If the end stud of the soffit ties into an outside corner, set the end stud to follow the frame line of the wall. The tapers will often have reached the room before you, so the outside corners are beaded and taped. When this happens, you're going to have to establish your finish elevation, and cut the corner bead off at that point. A hacksaw works well for this task. Cut through the tape with your utility knife to avoid pulling it loose. Line up the end stud with the back side of the drywall already hung on the wall, as shown in Figure 4-5.

Figure 4-5. The end stud for this drop is set 1/4 inch above the plate line and flush with the back side of drywall on the intersecting wall.

Use the Bottom Plate

In situations where only one end of a soffit framed to the deck ties into an already-framed and- rocked wall, use the bottom plate to plumb the soffit's outside corner stud. First cut the bottom plate to length, about $1/4$ inch back from the actual frame line of the corner. If the soffit is no longer than a single stick of plate, just cut the bottom plate to the same length as the top plate. On longer soffits, splice the plate together and plumb the corner point down to get your length.

First, securely fasten the bottom plate in place, set the soffit to a string line (commonly called a *dryline*), and kick it off. Then plumb the corner stud down from the top plate with your level. Once plumbed, screw the stud off in the plate to lock it in place. Use the bottom plate of the soffit and wall to hold the corner plumb from the other direction, overlapping the plate at the corner (Figure 4–6).

Figure 4–6. The bottom plate of the intersecting side of the soffit is cut to overlap. Once plumbed, the plate is screwed together with one framing screw, which locks this end in place while allowing the opposite end to pivot as it's plumbed.

Horizontal Control Stud

When framing extremely tall soffits and suspension walls (especially when the studs run wild), you'll need to hold the studs on layout at the bottom. To take care of this situation, use a horizontal control stud. A 25 gauge $3^{5/8}$-inch stud will do the job just fine. Screw the control stud to an end stud that's tied to a wall, or cut a shoe on one end and tie it into the wall itself (Figure 4–7). On long soffits or suspension walls, splice the control studs together the same way you splice plate together.

Run the control stud at about the middle of the studs and level it with your 4- or 6-foot level. Fasten the control stud to the inside of the work with panheads, where it'll be covered up. You can run the control stud on the outside of the work, as long as it's pulled back off as the wall is rocked up.

Figure 4–7. We used a horizontal control stud to hold these soffit studs, which have been run wild, plumb and on layout until the framework was complete. Then we took it off before the drop was rocked.

Plating the Bottom of the Studs

This method is adaptable to any suspended framing conditions, and will work the same whether all or some of the studs have been fastened in the top plate in advance. To begin, slide the end of the plate onto the end or corner stud of the soffit, and clamp it in place. Allow the stick of plate to run at an angle, just under the rest of the studs, and work your way down the wall, twisting each stud and raising the plate onto it, as shown in Figure 4–8. Clamp the plate to a stud every 4 feet to the end of the stick of plate, then establish and pull layout for the studs. Using the splicing technique, splice the bottom plate together and plate the studs to the end of the wall or soffit.

Kicking Off the Soffit

As you build the soffit, set the studs or jigs of the soffit to a dryline to keep the soffit uniformly plumb. Use kickers to hold the soffit studs or jigs to the dryline. In Chapter 1 we covered cutting the kickers. Here we'll put them to work. In most cases, these kickers will only need a shoe cut on one end, and you can eyeball the cut for the shoe. While the shoe doesn't *have* to be cut perfectly square, making cuts like these is a great time to practice making square cuts by eye. When kicking off a soffit, the kickers, like nearly all bracing and support material, are spaced 4 feet on center. As with all kickers, you'll get the maximum support they're capable of by running them at a 45-degree angle, as in Figure 4–9.

Keep this in mind as you determine the length of a kicker. There are also two other factors to remember. First, make sure the kicker will fasten to the stud 6 inches above ceiling height. Second, don't let the kicker stick out on the other side of the stud once it's plumb.

Plumbing a Soffit with a Dryline

To plumb the soffit studs or jigs in a uniform straight line, use a dryline. You'll want a good braided nylon dryline that's around 135 pound test. It has to stand up to being pulled extremely tight to give the dryline its accuracy. But even pulled tight, it's not really accurate when pulled much farther than 40 feet. On extremely long work, plumb and brace a stud every 40 or 50 feet and pull your dryline between them.

Set the Dryline

Your first step in pulling a dryline is to plumb and brace the two end studs by either tying them in to an end wall or kicking them off. Next, you'll cut two pieces of drywall (for gauge blocks) approximately 3 inches by 4 inches. Then screw one gauge block to each end stud, about 2 inches up from the bottom (Figure 4–10). Finally, add one more screw in the center of the drywall, leaving the head of the screw sticking about 1/4 inch out of the drywall. You can see it Figure 4–10.

Figure 4–8. As the plate is spliced and clamped at the joint, the studs are twisted and the plate is raised up onto them.

Figure 4–9. You can see the framing kicked off every 4 feet, plus a diagonal brace and horizontal control stud. This is common on tall suspension walls like this one, to make the soffit or suspension wall solid with as little work as necessary.

Figure 4–10. Here's the gauge block screwed to the end stud of the drop, with the dryline hooked to it.

Now you can hook the dryline to the screw left sticking out of the drywall. Tie a loop in the end of your dryline and slip it onto the protruding screw at one end of the soffit. Pull the dryline to the opposite end of the soffit, loop the dryline over your index finger and spin seven winds into the dryline before slipping it over the screw at that end. See Figure 4-11. Now pull the dryline tight by pulling the slack in the line running from screw to screw, while at the same time pulling on the string line spool, pulling the slack through the winds in the dryline. The winds act as a pulley and help hold the line tight while you regrip the line to pull out more slack (Figure 4-12). You must pull the dryline very tight. When it's tight, cinch the carpenter's knot down by pulling the spool of the dryline directly back to the screw, then around the screw a couple of times and set the spool in a cradle. You form the cradle the same as tying a knot, only you pull down the spool when it's half way through the loop.

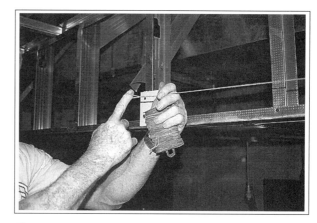

Figure 4-11. With the gauge block screwed to the end stud, the dryline is wound up to slip over the head of the screw that was left sticking out.

Setting the Studs to a Dryline

Beginning at either end, you'll set a stud to the dryline every 4 feet, bracing it in place with a kicker. Set the gap between the stud and the dryline the same thickness as the drywall you used to shim out the dryline at each end. Position the kicker against the hard side of the stud, eyeball the gap, and clamp the kicker to the stud. Check the gap and make any needed adjustments.

When setting studs, clamp the studs loosely to the kicker and adjust them by tapping the stud with your axe while checking the gap (Figure 4-13). Once you've set the stud, screw the stud and kicker together using two or three panheads. As you work your way down the wall, setting each 4-foot stud, the plate will pull the rest of the studs in line with them. Figure 4-14 shows a soffit set to a dryline and kicked off.

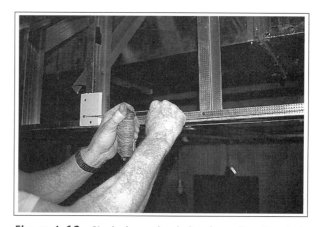

Figure 4-12. Cinch down the dryline by pulling the slack out, then pull the slack through the carpenter's knot.

Setting the Bottom Plate to a Chalk Line

As with other soffit methods we've looked at, this method will also work in many other framing situations. In this step of the soffit framing,

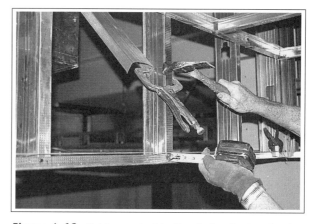

Figure 4-13. Setting a stud to the dryline by tapping it into position with a drywall hatchet while measuring the gap between the stud and dryline.

note the importance of cutting the studs 3/8 inch short and stuffing them tightly up into the top plate. Begin by establishing the plate line elevation at both ends of the soffit. You'll figure the elevation off of the floor, benchmarks, or ceiling when you're framing the soffit to a ceiling. Once you've established the bottom plate elevation, set the plate at each end, then measure up from the plate 3 inches on each end. Double-check the studs to ensure they're on layout, then snap a chalk line from end to end of the soffit. To snap the chalk line without help, run a framing screw into the 3-inch mark at one end of the soffit and hook the chalk line end to it. Remember, 50 feet is as far as you'll want to pull a chalk line. On work longer than 50 feet, establish elevation and make additional 3-inch marks, splitting the distance.

Double-Duty Dryline

Now that we've covered the common technique of setting the bottom plate to a 3-inch chalk line snapped across the bottom of the studs, let's look at a similar option. In many situations, you can eliminate this conventional method by setting the dryline exactly 3 inches up from the bottom frame line elevation. Set up the gauge blocks centered 3 inches above the bottom plate elevation, then screw them in place. Now measure up from the plate elevation 3 inches and mark the pieces of rock. Run in the screws you'll hook the dryline to at the 3-inch mark, and pull the line as we discussed before. Figure 4–15 shows a double-duty dryline.

You can now combine two steps: plumbing the studs with a dryline, and setting the bottom plate to a 3-inch chalk line. Study both methods for individual details. Combining both techniques will allow you to do in one pass the work that usually requires several passes back and forth across the soffit or suspension wall.

There's an alternate technique you can use once the plate or utility angle is slipped onto the studs. When the plate is on the studs, set the two ends. Next run a framing screw directly into the corner of the plate or angle at each end, and pull the dryline from screw to screw (Figure 4–16). Then set the edge of the plate to the dryline for both up-and-down and in-and-out. The drawback to this method is that if any part of the framing pushes the dryline, it'll throw off the rest of the framework.

Setting the Plate

With the chalk line snapped, you're ready to set the plate. First, adjust your clamps so they'll hold the plate in position but still allow you to adjust it up and down. Clamp off four or five

Figure 4–14. Here's a complex soffit set to a dryline and kicked off. The soffit is just as straight as it looks, braced every 4 feet with a kicker from the back side.

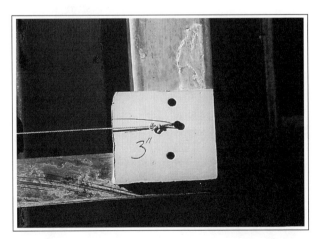

Figure 4–15. This is a double-duty dryline that's properly set up. The gauge block (made from scrap drywall) is screwed to the back of this end jig. An elevation mark is established 3 inches above the plate line. Then a fourth drywall screw is run in at that point, with the head sticking out. Set up another gauge block at the opposite end of the soffit, then pull a dryline between them. Now you can set the jigs for both plumb and elevation.

studs on layout at a time, and remove any framing screws used to tack the plate in place. Measure from the bottom of the plate to the 3-inch line at each stud, tapping the plate up and down with your axe as needed to set the plate at exactly 3 inches. See Figure 4–17.

Set all of the studs you've clamped and then screw off each of the studs with one screw per stud. Pull your clamps off and move on down the soffit, repeating this procedure to the end of the work. Figure 4–18 shows the bottom plate set and screwed off. To set the back side of the plate, either level the plate at each stud or level it at each end and snap another 3-inch line. In either case you'll again clamp four or five studs, set them all, then screw them off.

Figure 4–16. The framing screw runs right into the bottom corner of the plate. The loop at the end of the dryline slips over the head of the screw, while the other end loops over another screw at the opposite end of the bottom plate.

Building Jigs

Let's take a look at building jigs. But first let me quickly explain the difference between a jig and *jigs* as I use the terms. You'll find these terms commonly used in the trade.

- The *jig* is the mold that you put the cut pieces of stud into to hold them in place while you fasten them together.
- *Jigs* are the completed parts that you built in the jig.

You'll use jigs to build soffits as well as other metal stud framing that has multiple elevations or angles. Jigs help you maintain consistency as well as speed the building process when you're doing repetitive work. If you're a residential wood framer making the transition to commercial work, you'll recognize trusses as being very large jigs built on a very large jig table.

Building the jig is very precise work that requires a great deal of thought and experience. If the jig is flawed, the soffit you're building will also be flawed. There's a good deal of responsibility here. That's why the jig is usually built by the job site foreman or his right-hand man. But if you're ever called upon to build or help build a jig, this information will get you going in the right direction, plus give you a better understanding of how to put the jigs together.

Figure 4–17. The plate is clamped to the stud, where it can be tapped up or down until it's set to the 3-inch mark on the tape while measuring to the chalk line.

Figure 4–18. The bottom plate of this $3^{5}/_{8}$-inch suspension wall has been set to the chalk line.

The Jig

If you build the jig on a sheet of plywood, you can tightly screw down the pieces of plate the jig is made of, preventing any movement. Then set up the resulting jig table on a couple of sawhorses to eliminate bending and stooping to the work. You'll build the jig according to a detail of the soffit, found in the specification section of the prints. There will also be variables not shown in the prints that you'll need to consider. Check how each jig will tie into the rest of the framing, and look for any changes made since the prints were drawn up. You'll also need to allow for the thickness of the drywall that'll be used to rock the soffit. The prints will show the "finish" dimensions only. You'll have to refer to the wall legend to find out the thickness of the rock to be used.

Once you have the jig figured out, lay it out on a sheet of plywood, beginning with the longest section of the jig. For each section of the jig, lay out both sides of the material to be used ($3^{5}/_{8}$-inch material is the most common). Use your framing square to form all corners, both inside and out. Once you have the jig laid out, establish any "stop" points for sections of the jigs that will require an exact length.

Cut the pieces of plate you use to form the jig to length, allowing $^{1}/_{4}$ inch off the total length of all the pieces of plate. Then cut the inside leg off at all the corners to overlap the plate. You can cut the stops out of either plate or stud material. The stop's only job is to give certain sections of the jigs something to bump up against, setting them to a specific length.

Overlap the corners of the cut pieces of plate and screw them down to the plywood with drywall screws spaced every 8 inches on both outside edges and at the ends of each piece of plate. Fasten the jig pieces together hard side to hard side. To do this, you'll have to elevate the intersecting sections of the jig by adding a stud under the plate. Screw the stud off solid in the plate, then replate that section of the jig on top of the stud (Figure 4–19). This will allow the sections of the jigs to overlap and be screwed together.

Building the Jigs

There's an obvious order to putting the studs in the jig. Install the bottom pieces (as they sit in the jig) first, hard side up. These are usually the longer sections of the jig parts. Next install the intersecting sections hard side down in the upper sections of the jig. Slide all the sections with stops up against the stops, and push the pieces down tightly into the jig. Now screw the joints together, with one screw in each of the four corners of the joints (Figure 4–20).

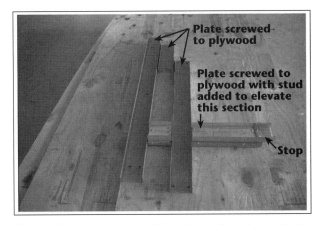

Figure 4–19. A completed jig with the lengths marked for each leg of the jig parts. There's a piece of stud added to the end on one section of the jig as a stop.

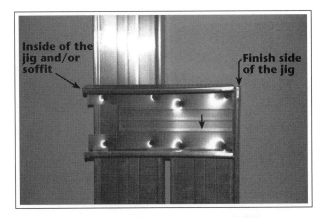

Figure 4–20. Here, two joints side by side in the jig have been screwed off with four framing screws in each. Notice the jig parts are held back from the outside corner point on the finish side of the jig. Never let the ends of the jig parts overhang each other there. Notice also that the jig part overhangs at the inside of the jig. This is OK as long as it isn't a finish section of the soffit.

Measure the lengths for each of the sections and write the length in the corresponding section of the jig. Now cut the pieces for and build one of the jigs. Check it for square and dimensions, then clamp it in place as the jigs will be framed. Check the fit to make sure everything works out, and you're ready to go.

Next, figure out the number of jigs you'll need for the layout. It's common for large projects to use several hundred jigs. If that's the case, you'll need to team up. Have one carpenter cut while the other builds the jigs. Measure the lengths for each of the sections and write the length in the corresponding section of the jig.

Cutting the pieces to length is very time-consuming. The fastest way to cut your material is with a chop saw. Build a stop on the chop saw table for the material to bump into, eliminating the need to measure each piece. Set up a cut table with sawhorses and plywood, and position the chop saw where you want it. Then cut pieces of stud and screw them down tightly around the chop saw, locking the saw in place. From the chop saw blade, measure and mark the various lengths of material needed on the sheet of plywood, square to the blade.

Now cut a piece of 6-inch stud 12 inches long to use as a stop. Then set the stop on edge with the end of the stud even with the length mark of the length you're cutting, and screw it down to the sheet of plywood (Figure 4–21). It will obviously be important that you cut the 6-inch stud square, and check the length once it's screwed in place. Recheck it each morning as you set up. You'll also want to check a piece for length after every 20 cuts to ensure they're accurate.

When cutting material for sections of the jigs that need tabs or one of the legs cut off, cut about 10 or 20 pieces to length, then make the additional cuts. Stack your cut material in neat stacks according to their lengths. Also neatly stack the jigs as they're built.

In Figure 4–22, you see a first rate jig operation set up and working. This work can get monotonous, but take pride in your work. The better your jigs, the smoother the job will go. Quality jigs make the framing easy.

Figure 4–21. The cut table is set up on planks. The stop is cut from 6-inch plate and screwed down on edge.

Framing with Jigs

Once you've finished building the jigs, there are a couple of ways to frame them in—after you've established the soffit elevation, of course. One way to fasten the jigs in place is to screw them to the studs of a wall that's already been framed. The other way to frame the jigs is to fasten them into a top plate. Let's take a look at each of these methods separately.

Framing to Wall Studs

If you're framing the jigs to the studs of an already-framed wall, build the jigs with two fastening points, usually at the top and bottom points of the jigs. Keep this in mind as you design and build the jig. The two sections of the jigs will fasten to the studs hard side to hard side. When you build the jig table, the direction the jig faces as it's screwed down to the plywood will determine which direction the hard side of the jigs will face — right or left.

Your first step is to establish the elevation for both the top and bottom fastening points of the jigs at each end of the soffit, then snap a line for each elevation. In most situations, the ends of the soffit will tie into an intersecting wall or column, making it easy to set and tie in the ends of the soffit. See Figure 4–23. You'll frequently run into situations where the jigs can't be tied in on one end of the soffit because of the way the jigs are built. When that happens, lay out the jigs on the end wall. Then cut individual sections of the jigs and tie them in to the end wall or column,

Figure 4–22. A three-man crew is building the jig parts. Over by the gang boxes one carpenter is cutting the studs to length, as the other two are putting the jig parts together. Overhead you can see the jigged soffit going up.

as in Figure 4-24. Once your ends are set exactly to the elevation and distance off of the wall studs and plumbed, pull a dryline from end to end.

In cases where the end jigs don't tie into a wall or other framing, you'll use the same procedure — up to the point of setting the jigs for elevation, distance off of the wall, and plumb. The difference is how you brace the end jigs so that the tension of the dryline won't pull them out of square. Remember, for every action, there's an equal and opposite reaction. If the jigs pull out of square, the dryline will pull in closer to the wall. That leads to framing the entire soffit short of the actual plate line. To overcome this problem, run a temporary horizontal kicker to the bottom of the end jigs, holding them square against the tension of the dryline (Figure 4-25). As you can see in the photo, the kicker must be fastened to the end jig at the outside point near where the dryline is attached to the jig.

Setting the Jigs

Let's assume that the drywall you used to shim out the dryline is $5/8$ inch. Set the jigs against the studs hard side to hard side with the bottom of the jigs even with your bottom elevation line (chalk line). At the same time, eyeball the front of the jigs as close as you can to $5/8$ inch off of the dryline and clamp the jig in place at the bottom fastening point. Pivot the jig so that the top fastening section of the jig lines up with the top elevation line, and clamp it to the wall stud. Then adjust the jigs to the dryline until there's a $5/8$-inch gap between the jigs and the dryline. While you're setting the jigs to the dryline, you must also maintain the elevation. If the jigs are lined up with both elevation lines and $5/8$ inch off the dryline, they should be plumb.

Set the first two or three jigs using the three points (two elevation lines and the dryline) and check them for plumb. If it's working, keep

Chapter 4: Soffit Methods   67

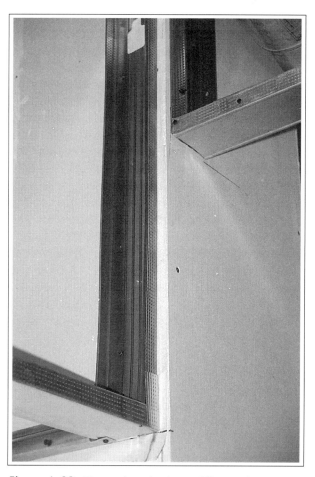

Figure 4–23. The end studs of two different elevations of an intricate drop are tied into an already-framed and rocked column.

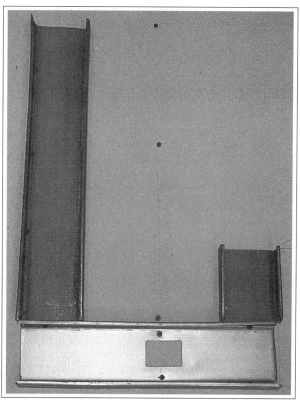

Figure 4–24. Pieces of 3⁵⁄₈-inch stud cut to form this light trough are individually tied into this end wall. The individual jig sections are tied into the wall by both screwing through the drywall and into the wall studs, and by cross-tying the studs to the drywall.

going. Check every fifth jig with your level for plumb to make sure it keeps working. When adjusting the jigs for plumb, only pop the top clamp loose to make your adjustment. Then use the clamp as pliers to control how much the jigs move. The bottom clamp will maintain the elevation. As you make the adjustment for plumb, double-check your elevation and dryline. When all the points are on, screw it off at both the top and bottom, fastening sections of the jigs with three framing screws in a triangle pattern. Place two screws close to the edge of the wall stud and one toward the back edge.

When using this method, the layout of the jigs will follow the layout of the wall studs, which is usually 16 inches on center. See Figure 4–26.

Figure 4–25. This jig, which is being framed to a suspension wall, has been kicked off with a horizontal kicker to hold the jig square against the tension of the dryline.

Figure 4–26. These large truss jigs have been set to a chalk line to keep it all straight. The jigs, once set, are screwed off to the wall studs hard side to hard side.

Figure 4–27. Several jigs in the top plate as they're set to the laser beam. The laser is fastened to the wall studs at the far end of the drop.

Setting Jigs in a Top Plate

Assuming your top plate has been laid out and screwed or shot up, we'll establish the elevation first. Again, using a laser will greatly simplify the job. Clamp each of the jigs in the top plate, then measure from the center of the beam to the bottom of the jigs. You can see that being done in Figure 4–27. As you check the elevation, it's important for the jigs to be plumb. Instead of trying to plumb the jigs with a level, swing the jigs slowly back and forth perpendicular to the top plate, while measuring from the jigs to the laser beam. As the jigs swing past the widest measurement, the jigs are plumb. This is your true reading. To make adjustments, pop the clamp loose and move the jigs up or down. The light pressure of the clamp will allow you to make fine adjustments a lot easier. Once you've got the elevation set, screw the jigs to the top plate with a single screw in both the front and back of the plate.

When framing jigs to a top plate, you'll often be building the soffit to an already-framed suspended drywall ceiling (Figure 4–28). When the elevation is already established by the ceiling, simply measure down from the ceiling to the bottom of the end (or corner) jigs. It's common to set all the jigs to the ceiling, but any unnoticed flaws in the ceiling will stand out in your soffit. When in doubt, plumb and kick off the end jigs and pull a dryline. Again, run a temporary kicker to hold against the tension of the dryline. A double-duty dryline will save you lots of time by letting you set each jig's elevation while you plumb them to the dryline and kick them off, all in one pass of the work.

Plumbing the Jigs

You'll set the jigs plumb to a dryline, and hold them in place with kickers shot or screwed to a solid part of the building. Line up the kickers squarely with the hard side of the jigs, every 4 feet. With the kickers shot in place, eyeball the gap between the jigs and dryline and clamp the jigs to the kickers. Attach the kickers to the jigs above ceiling height, making sure that the end of the kicker doesn't stick out past the opposite side of the jigs. As you measure the gap and make adjustments, don't take the clamp off. Just pop it loose and use hand pressure to control how much the jig moves. Once the jigs are set, screw the jigs and kickers together with two screws. The plate and/or utility angle connecting the jigs will pull the rest of the layout jigs in line. See Figure 4–28.

Patterns

Let's take a look at patterns and best way to lay them out and cut them. You can use patterns cut from either drywall or plywood to form radius soffits and walls. Once the patterns are cut, use them to mark similar pieces of drywall or plywood, which you'll cut and screw to the radius plate to form the curved top and bottom plate of radius walls and soffits. In the field it's common for all the curved pieces of material to be referred to as *patterns,* with the actual patterns clearly marked "PAT." To keep things simple and clear in our discussion, I'll refer to the pieces laid out with the patterns as *templates.*

Before you choose the type of material to cut the templates from, consider the way the radius plate you'll form with the templates will be fastened in place. When the radius plate will be shot in place with a powder-actuated nail set, use plywood templates to form the plate. The pins would simply blow through both the plate and templates if you used drywall. Also use plywood to form the radius plate whenever the framing will carry the weight of the framing and drywall. Since drywall templates are much faster and easier to cut, I recommend using them instead of plywood whenever these conditions aren't present.

Lay Out and Cut the Pattern

To form the patterns, you've got to clear and sweep an area large enough to strike the arc of the radius. Whenever possible, lay out and strike the arc directly under where the work will be framed. This lets you swing the arc on the floor and later plumb up the radius. While this is the ideal, you'll frame many radius soffits that are much larger than the area they're framed into. For example, Figure 4–29 shows a 75-foot radius canopy soffit built 20 feet off of a wall. For this soffit we had to set out the plywood in the parking lot, where we had enough room to swing the arc and mark the patterns.

Figure 4–28. This carpenter has four jigs screwed off in the top plate, all connected with utility angle. With these jigs kicked off and set to a dryline, the rest of the jigs in this light cove can now be set. They'll simply follow the utility angle and jigs that are kicked off.

Figure 4–29. This radius canopy was so large that to form it we laid sheets of plywood end to end out in the parking lot and struck the arc for both sides of the plate on them. The entire soffit is supported with lookouts.

To establish the center point of the arc, you'll usually pull (measure) off of two intersecting walls. In some cases, however, you can simply mark them where there's enough room to swing the arc. First mark the center point, then drive a concrete pin into it with your hammer. You can also use an angle nail (hard nail) used by the grid ceiling men. In either case, the pin or nail must be solid.

From the pin, measure out the radius, then add or subtract for the thickness of the plate, and allow for the drywall. If you're striking an inside arc, add the width of the plate. When swinging an outside arc, subtract the thickness of the plate. Allow for the drywall in either case. Swing the arc for both sides of the plate, marking the radius on the concrete as a reference line as well as for making sure the arc is smooth.

Now lay your pattern material across the arc. Swing both sides of the arc across your pattern material, as shown in Figure 4-30. Once the pattern is laid out, cut it out. If you're using drywall, score the rock along the layout lines, then cut it with a router. Use a jigsaw on plywood, but in either case make sure the cuts are smooth. Then clearly mark the pattern "PAT." You can keep the outside of the radius left on the scrap drywall to help form the radius at the joints as the soffit is framed.

The Templates and Plate

Cut the templates from the same length material as the patterns, so you can line up the ends. Lining up the square factory ends of the pattern with the factory ends of the template material is extremely important. The square ends of the templates will later butt tightly together, helping to form the smooth curve of the radius. Lay the pattern across the sheet of drywall and line up the ends squarely, then trace around the pattern (Figure 4-31). Keep your pencil sharp to help keep the pencil line tight to the pattern. As you lay out the templates, allow a 1-inch gap in between each template. Cut out the templates following the same procedure you used to cut the pattern.

Screw off the radius plate (covered back in Chapter 1) to the templates using drywall screws. Stagger them from front to back of the plate, spaced 6 inches apart. The plate will also need to be cut to length. Make your cut after you've got the plate screwed off to the template. Line up one end of the plate with the end of the template and screw off the plate. Start at one end and work your way around to the opposite end. Clamp the plate to the template every 6 inches or so (Figure 4-32). At this end of the template, cut the plate the rest of the way through one of the relief cuts so the plate is even with the end of the template.

When framing the radius plate to the deck, plumb up the radius plate line in two places for each template section. Then set the plate and

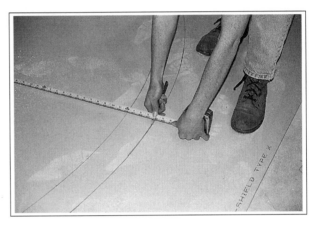

Figure 4-30. Striking an arc for both sides of the plate onto a sheet of $5/8$-inch drywall. As you can see, Angie is keeping her lines smooth and round. Remember the drop will be as smooth as the arc you strike.

Figure 4-31. Using the pattern, Kelly is tracing out the templates on a sheet of $5/8$-inch drywall. Once all the templates are traced out on the sheet, they'll all be cut out at once using a router.

templates to the plumb marks which will form the radius. To form the bottom plate of radius soffits and suspension walls, as well as the top and bottom plate of protrusion framing (lookouts), add a second layer of templates. Stagger the joints between the different layers of templates by half to form a smooth radius. When using two layers of templates, make sure you allow for the second layer of template material on the stud length. The second layer of templates is often added after the studs are stuffed.

Figure 4–32. The plate clamped to the drywall template, with the screws run into the plate right at the clamps to keep the plate from pushing away.

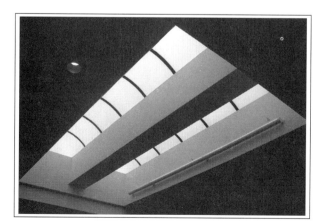

Figure 4–33. An I-beam runs right through this skylight. Lookouts were used to support the framing and drywall that now encases it.

Supporting Soffits

On most jobs, supporting a soffit will be as simple as shooting up a top plate and screwing the studs to it. But if this is the only thing you know how to do, sooner or later you'll trip over a variable you don't know how to handle. That's why we'll take a look at a couple of other methods used to support a soffit in not-so-common conditions. We've already covered how to use CRC and suspension wires to support the framing, including soffits. So let's move on to some alternative ways to support a soffit.

Lookouts

Lookouts are another way to support a soffit. They'll carry a soffit built around members of the red iron superstructure. Lookouts are commonly called *protrusion framing*. This technique is useful for framing exterior canopies and skylights. You'll also put this method to work for framing columns, using the lookouts to carry the top plate. The lookouts let you build a solid soffit without using suspension wires, kickers or any other outside bracing. All the support is inside the soffit. For this example, we'll focus on a soffit built around a section of the superstructure running through a skylight, like the ones in Figure 4–33.

Establish Layout

As with all soffit work, the surrounding framework must be completed first in order to get the soffit properly positioned. In this case, that would be the perimeter walls of the skylight. The layout work is much faster and easier if the skylight is already rocked. Soffits that run through a skylight like this will nearly always be centered in the skylight, so the first step is to establish the center point of the skylight at both ends of the red iron. From the center point, measure half the frame width of the soffit each way, and plumb the points with your level. Draw the frame line well past the top and bottom elevations.

Your actual layout for the soffit will require a bit more work than this, but for the lookouts all you'll be concerned with is the width of your soffit. Once in a blue moon, the elevation of the soffit will dictate what width of stud you'll cut the lookouts from. Obviously, if the soffit frame line elevation is only 4 inches below the I-beam, you can't cut the lookouts from 6-inch material.

But the material width is rarely an issue. Most lookouts are cut from 3⁵⁄₈-inch stud material. But watching for this condition will help keep you out of trouble.

Measure between the frame lines to get the length of the lookouts. Then, assuming the soffit studs will also be cut from 3⁵⁄₈-inch material, subtract 2 inches for play. If the soffit studs were 1⁵⁄₈ inches, you don't need to allow that much for play. One-half inch on each side will be plenty. You would just have to be more careful as the lookouts are shot in place. The lookouts will be spaced 48 inches apart on both the top and bottom of the red iron. Keep this in mind as you cut all the needed lookouts to length.

Pull Layout

The layout for the lookouts will follow the layout of the soffit studs. The soffit studs will be screwed to the lookouts hard side to hard side. Since you know you'll have a layout stud at 48 inches, add ³⁄₄ inch (half the width of the studs) and mark the top of the red iron at 48³⁄₄ inches. This is assuming the skylight is rocked. If it isn't, you'll need to also add the thickness of the drywall. Next, square the layout mark across the red iron, because the red iron is rarely square in relationship with the skylight framing. It's best to mark two points at 48³⁄₄ inches and use a straightedge between them. Pull layout for the hard side of the lookouts, so mark the far side of your layout with an X. Pull the layout for the rest of the lookouts from here, marking every 48 inches. Using this method, lay out the bottom of the red iron also. As you're pulling layout, your partner will be cutting the lookouts to length.

Shooting the Lookouts in Place

As you prepare to shoot the lookouts in place, you'll need to determine how far the lookouts will stick out past the red iron. First, measure from the red iron to the soffit frame line you plumbed earlier at both ends of the red iron. You need to see how far the red iron is out of square with the skylight walls. If the difference is ³⁄₄ inch or more, you'll need to set the end studs of the soffit, tie them into the skylight walls and pull a dryline, setting each lookout to the dryline. In most cases you can set the lookouts using the red iron itself. Set a lookout on top of the red iron against the end wall and slide the end of the lookout 1 inch from the frame line of the soffit, centering the lookout inside the soffit. Then measure from the edge of the red iron to the end of the lookout, and use this number to lay out each lookout.

Keep in mind that the lookouts must be marked differently for the top and bottom of the red iron to set all the lookouts to the same side of the red iron. Draw the reference mark up around the bend in the stud on the soft side. That'll make it easier to see as you shoot the lookouts in place.

Set the lookouts on edge against the red iron and fasten them in place with a powder-actuated nail set. Getting the shotgun into position will take a little effort. Set the lookout on layout and line up the reference mark with the side of the red iron. Then set the end of the shotgun to the inside leg of the lookout and push down as you move the shotgun into a plumb position, as shown in Figure 4-34. The pressure of pushing the barrel down into firing position will help hold the lookout in place. As you stand the shotgun in plumb position, the light gauge stud will fold back out of the way. Shoot each lookout in place with two steel pins, using yellow loads. Figure 4-35 shows the lookouts shot in place, with a few soffit studs fastened to them.

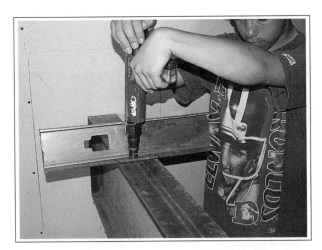

Figure 4-34. You can see the shotgun in position, forcing the 25 gauge lookout (stud) back to allow access. If you look closely, you can see the soffit frame line laid out on the end wall. The lookout itself also has layout marks to center it up inside the framework of the soffit.

Safety glasses are a must. Even under the best conditions, the pins have a tendency to shear off, sending part of the pin as well as small fragments flying. I also recommend that you use of some type of ear protection.

Using the Drywall to Carry the Soffit

When a sheet of drywall is laid flat, it's weak, flimsy and easy to break. Turn the sheet on edge and keep it straight and it's rigid and strong. You can use this phenomenon on small soffits that can be spanned with a single sheet. It lets you build casework soffits quickly. I won't go into detail for this method. Simply knowing that this is a valid means of supporting a small soffit is as in-depth as we need to go for now. In Chapter 6, we'll be putting this technique to work, so we'll cover all the details there. This trick is a big timesaver.

As we close out the *Tricks of the Trade* section of the book, keep in mind that each of these methods will be put to work in many framing situations. Just like a moto-cross track, commercial metal stud framing has fast straightaways combined with bumps, jumps, and hairpin turns. Keep all of these tricks in mind, and put them to work in any situation where you can save time and material.

Figure 4–35. A drop being framed around an I-beam using lookouts. Notice that Larry has simply laid several of the lookouts down with the hard side to the I-beam. Larry's using 14 gauge plate, which is rigid enough for this little drop.

SECTION 11
Step-by-Step Methods

In the first four chapters, we looked at tricks of the trade that'll help make your framing jobs run smoothly and quickly. Now it's time to focus on the step-by-step process of building walls, ceilings and soffits. I'll walk you through the process of framing the work, with side trips into some of the complications you'll run into on the job.

If you're a residential wood framer, you're accustomed to having a job pretty much to yourself until you're finished. That's about to change. You'll have a lot of adjustments to make, including sometimes having a wide range of tradesmen and their material underfoot. Sometimes you'll have other trades working the wall you're framing just as fast as you get the studs stood up and screwed off. A good dose of tact, mixed with some negotiation skills, will get you a lot farther than a quick temper. Confrontation generally leads to more problems than you started out with. When you do have the job to yourself, you'll have clear sailing to frame up linear feet of walls that just can't be equaled in wood framing.

If you're a rookie using this book to get going in the trades, I suggest you pay attention to two factors. You need to learn the procedures, of course, but also take note of the work process. Being fast isn't enough — you have to do quality work at the same time.

This book alone won't make you a journeyman, but it will greatly reduce the time it takes you to achieve journeyman's status. With this book, some experience, and with tips you pick up from those you work with, you'll soon have all the skills you need to do the job efficiently and effectively. When I started framing with steel, there was very little information around. I learned everything by doing it and making the mistakes, and from other metal stud framers who had learned the same way. But I've been lucky enough to work with some of the top hands in the country. As a rookie or a journeyman, keep an open mind and you'll never stop learning.

While everyone has to pay his dues, know that in many parts of the country, a journeyman metal stud framer can earn well over $20.00 an hour. Many large outfits also offer health benefits, paid vacations, and 401(k) plans, as well as profit sharing and bonuses. Paying your dues is a small price for what you can gain from a career in metal stud framing.

Chapter 5

Interior Walls

In this chapter, we'll take a look at the materials and procedures you'll use in framing interior light gauge metal stud walls. To begin with, there are seven basic widths of light gauge studs: $1^{5}/_{8}$, $2^{1}/_{2}$, $3^{5}/_{8}$, 4, 6, 8, and 10 inches. The most common lengths are 8, 10, 12, 14, 16, and 18 feet. Light gauge studs range from 25 gauge (the lightest), to 22 gauge and 20 gauge.

Before we start looking at building techniques for walls, there are three general cautions I want to share:

- First, keep in mind that quality should come before quantity. Haste *does* make waste in the form of rework, and there are no profits in rework. Quality — the mark of a true craftsman — is a learned trait. Quantity comes with experience.
- Second, you'll also need to learn to plan ahead for the length of stud you use for a particular wall. Estimators commonly bid jobs based on using the scrap cut from a length of stud for another wall or soffit. Keep this in mind as you make your cuts. Minimizing waste will maximize profit on the job.

- Finally, remember there's some preparatory work before you can start building walls on any job. That preliminary work includes these steps:

  1. Planning
  2. Clearing
  3. Cleaning
  4. Stocking material

Planning includes studying the prints for the area you're working in, and if possible going over the area with your foreman. Most foremen will walk through an area with you and get you "lined out"—instructed about the job.

Get the area cleared out. Too often, other trades stock material wherever it's convenient, with no regard for wall lines. This includes the drywall material that's often stocked either before or as the framing is begun. You (or your foreman) need to consult with the foreman of the crew the materials belong to, and let that crew move it. But don't get an attitude. You'll likely work with these people for many months. Just be assertive and let them know what needs to be done. Of course, some material will have to stay in the area. In this case, have it moved off the wall lines into a space where you can work around it. Watch where you have the material stocked, though. Don't stock materials in a room where it'll be difficult (or impossible) to get it out later.

Finally, get the area swept and stocked with your plate and studs. Put your materials close to where they'll be used. Most outfits will have a crew of laborers whose job is to get the area prepped and stocked for you. I make it a point to get to know my laborers straight off. They're the backbone of the crew, so treat them with respect. Get your cords rolled out, and your chop saw set up. You'll also want to gather up your scaffold and get it built. When all the prep work is done, you start framing, and keep framing without interruption. A good pair of partners will split these responsibilities. One works with the foreman while the other works with the laborers getting the area prepped.

We'll cover three basic types of walls here (with a side excursion into arches). While each type of wall has variations, these fundamentals should give you the knowledge and confidence to handle the simple as well as the complicated conditions you'll run into. Figure 5–1 shows the three kinds of walls: walls to the deck, freestanding walls and chase walls.

Walls to the Deck

With your work area cleared, stocked, and your equipment set up, let's get to it.

Step 1: Shooting Down the Bottom Plate Before you begin this step, make sure your wall lines are legible. Using your chalk box, resnap any lines that need it. The wall lines will either have both sides of the wall snapped out, or an *X* on one side of a single chalk line to tell which side of the line the wall will sit on. And there are two ways you can fasten down the plate. You'll usually shoot the plate down using a gas- or powder-actuated nail set. In rare situations (like old, brittle concrete), you may screw it down using a tap-con system (a hammer drill). With either method, space the pins or concrete screws approximately 24 inches apart.

Plating is a team job. The first partner cuts the plate, splices it together and lays it along the wall line. The other follows behind, shooting the plate down (Figure 5–2). Don't shoot the plate down so close to the end that the plate can't be spliced together. Once the two ends are spliced together, use one pin to hold the ends of both sticks of plate. As your partner cuts and lays the plate, it's not going to fall right on the wall line. You'll need to set the plate as you shoot it down. As you work your way down the plate, just tap the plate in place with your foot, setting the edge of the plate just on top of the chalk line. Then shoot it down. The plate must follow the chalk line exactly or you'll be framing dips and bumps into the walls.

Working your way down the wall, you'll run into windows, doors and other rough openings that you'll need to deal with. The windows won't affect the bottom plate. But when they're laid out along the wall line, you'll have to transfer the layout marks and the elevation for the window jamb to the plate once it's shot down. You should note that the door and window jambs have no real rough opening. The jambs fasten to the studs and they're a done deal. The door jambs are set in advance in some cases, but are

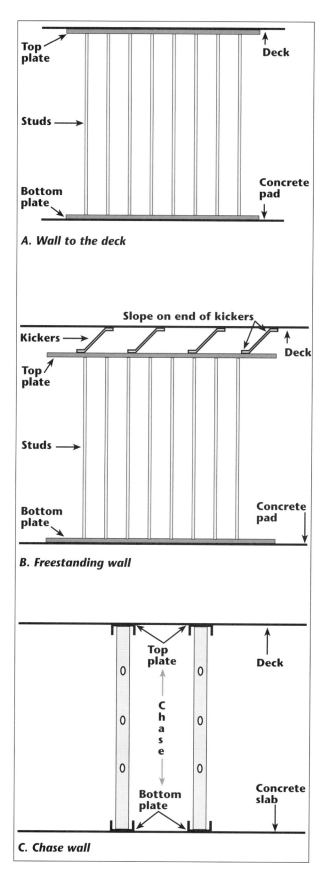

Figure 5–1. Three basic interior walls

Chapter 5: Interior Walls 79

usually set as the walls are framed. If the jamb is set, run the plate right up to the clip, as shown in Figure 5–3. If the door jamb isn't set, it will be laid out in advance. You'll have to stop the plate there and start again at the layout points for each side of the jamb. All other openings, like hallway entries, will also stop and start according to the layout marks along the wall line.

There are a couple of different ways to form a corner. We covered those methods in detail in Chapter 1. Always leave a ³/₄-inch gap between

Figure 5–2. Angie cuts the next stick of bottom plate as Kelly shoots down the first stick of plate in this wall.

Figure 5–3. The bottom plate of this 6–inch wall has been run right up to the door cleat.

the plate of intersecting walls at all inside corners. This gap allows the drywallers to slide the drywall of one wall inside the framing of the intersecting wall as they rock the walls. Then they'll tie the walls together with a *slap stud* (or *slider*).

Once you have all the plate cut and laid for the area, choose the walls that you'll be building first. It will usually be the long perimeter walls. Then pull the stud layout along the bottom plate, while your partner finishes up shooting down the plate.

You'll also have to watch for obstacles while shooting down the bottom plate, especially on multilevel projects like apartment and office buildings. On projects like these, the electricians, plumbers and tinners (HVAC installers) will run their lines up through the floor from level to level. The water lines and conduit will run right through your wall lines, as you can see in Figure 5–4.

Make a fairly tight notch in the plate around the conduit or water lines. The HVAC lines aren't usually in the wall line, but if the tinner gets his ductwork positioned wrong, it can cause some problems. Figure 5–5 shows this situation, and one way to deal with it. Stop the 6-inch plate on both sides of the obstacle and use $1^{5}/_{8}$-inch plate to run past it. Then cut a $1^{5}/_{8}$-inch header to run between the two closest 6-inch layout studs above ceiling height, and fasten $1^{5}/_{8}$-inch studs in place on layout.

You'll also run into walls that will have two or more plate lines even when there's only one common finish wall line. It usually happens in hallways, where portions of the wall may have large pipes running through them. Part of the hallway is framed with 6-inch material while the rest is $3^{5}/_{8}$-inch material. In this situation, both sizes of plate will follow the same frame line on the hallway side of the wall. The offsets for the different widths of plate will break up the partition walls of the individual rooms on the opposite side of the wall.

It gets a little more complicated when the various rooms on the opposite side of the hallway require different fire ratings. In that case, you may frame the hallway wall either with the same width material, or two or more different widths of material. It depends on what the *wall legend* calls for. The important issue is that the portions of the wall with longer burn rates

Figure 5–4. The bottom plate for several walls has been shot down, with conduit penetrating them all. The wall in the foreground shows the plate tightly notched around the conduit.

Figure 5–5. Here the wall switched to $1^{5}/_{8}$-inch material to get past an obstacle that came through the deck in the wall line.

require more layers of drywall while maintaining one finish wall line. To maintain that finish wall line, offset the framing of the multilayer portions of the wall back from the original frame line to allow for the additional layers of drywall (Figure 5–6).

First, you'll establish where the double or even triple layers of drywall start and stop. Then allow for the thickness of the additional layers and snap the offset frame line for this portion of the wall. Ideally, this layout work will be done in advance by your foreman or his layout man. But if it's not, no problem. You'll have the situation under control.

With the offset plate lines established, shoot the plate down, following the offset wall lines. Then pull the stud layout, maintaining one continuous layout through all the offset walls.

Figure 5–6. This is a good example of one wall line with different fire ratings. The outside (right side) of this wall will form one smooth finish wall; the inside is broken up by a partition wall that's been snapped out but not plated. The plates are offset by 5/8 inch. The wall in the top of the photo gets two layers, the second of which will carry on down the single-layer wall.

Step 2: Plumbing Up the Wall Lines As you prepare to plumb up the wall lines, think about where you'll want to start working. It's best to begin the framing at the perimeter and hallway walls. When these walls are done, move to the innermost point of the work area and build your way back out of the area. As you plumb up each wall, pop a line, then plate the deck and stuff it up (install the studs) while you're there. This will not only allow you to complete the work as you go, it'll also save a lot of wasted time spent tearing down and rebuilding scaffold.

In Chapter 1 we covered some important techniques used in plumbing up wall lines. The first time you're working the tops using a 16-ounce plumb bob with a partner on the ground who can't quickly steady the bob and guide you to the plumb position, you'll appreciate the importance of these tricks of the trade. As mentioned in the same Tricks of the Trade discussion, as you plumb up each end of the wall, you'll mark a plumb point in both directions, one plumb mark with the wall line, and the other 90 degrees to it. Use the 90-degree plumb mark to plumb up the stud layout, and to plumb up corner and intersecting wall measurements.

Use the same technique if you're using a pocket laser.

You'll run into a lot of obstacles while plumbing up and snapping the top plate wall line. Conduit, water lines and duct work are all often run under the deck. The ductwork and water lines will almost always be insulated, so there's usually room for the chalk line to pass over them. The conduit and vent lines, however, are often fastened tight to the deck with clips. Then you'll have to use your axe to pry conduit off the deck enough to pull your chalk line over it (Figure 5–7). This will help eliminate the need to plumb up on both sides of the conduit, which is the only other option. But for many obstacles, there won't be a choice. Some wall lines you'll have to plumb up in many places—bar joists, for example. For walls that are perpendicular to the bar joist and must run to the deck, you'll have to plumb up twice in between each set of bar joists. There's no need to chalk these. Just cut your plate to length and set it to the plumb points. For the sake of getting the walls up faster, your foreman will often get permission from the general contractor's superintendent to plate to the bottom of the bar joist.

To snap the top plate wall line, your partner will need another scaffold or a ladder to get up to the deck. To snap the top plate line by yourself, drive a concrete pin into the deck at one end of the wall and hook the end of your chalk line to it. On hollow metal decks, you can do the same thing with a framing screw. On the bar joist, clamp the end of your chalk line to the plumb mark. Mark an X on the side of the wall

Figure 5–7. Using a drywall axe to hold down a plumbing vent pipe, allowing the top plate line to be snapped out uninterrupted.

line that the plate will sit on, and you're ready to shoot the plate up.

Step 3: Shooting Up the Top Plate Fasten the first stick in place by yourself. Just hold the stick of plate in the middle and slide it to within 1/4 inch of the outside corner point. The end of the plate doesn't form the outside corner; the face of the end stud does—so there's a little room for play. Line up the edge of the plate on the chalk line as your partner eyeballs the corner point. Now shoot in a pin right next to the point where you have the edge of the plate set to the chalk line. You'll then work from end to end shooting up the plate, again spacing the pins 24 inches apart.

Using the splice to hold up one end of the top plate, plate the remainder of the deck alone, as shown in Figure 5–8. Your partner will roll your scaffold and keep you supplied with plate. When I'm working the ground, I'll cut my splices in the ends of the plate while it's still in the bundle. This way, you cut two pieces of plate at a time on all but the top and bottom stick of plate in the bundle. Hand up four or five sticks of plate at a time, and use any spare time to lay out the bottom plate and start getting the studs cut to length.

Many framers pull layout and stuff the studs on each stick of top plate as it's shot up. This method completes the wall with one pass. Others prefer to plate the deck, then go back and stuff it up. Every framer has their own technique. You'll soon figure out which method works best for you. Using either method, run the top plate continuous for the length of the wall.

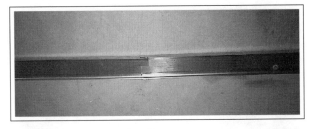

Figure 5–8. Here you see the top plate spliced together. First use the splice to hold up this end of the top plate as the other end is tacked up, then use the splice to fasten the ends of both sticks of plate in place with one pin.

Step 4: Lay Out the Plate A quality layout for the studs is essential for the drywall work to go quickly and smoothly. Establish the stud layout at the bottom plate, pulling it off of what the first sheet of drywall will butt up against, like intersecting walls or window mullions. Allow for the drywall when pulling the layout from intersecting walls, whether it's an inside or outside corner. Then pull the layout from this point, marking either 16 or 24 inches on center. The wall legend on the plans will call out the spacing you use. Using a pony clamp to hold the end of your tape in place lets you pull layout by yourself.

Once you've established the layout along the bottom plate, use the 90-degree plumb mark to transfer the stud layout to the top plate. Maintain an equal measurement between the plumb mark and the layout mark at the top and bottom plate to ensure that the stud layout is plumb. Pull layout along the top plate as far as you can reach from the scaffold, then get the studs stuffed and fastened in place before you move on. Pull the rest of the layout from the soft side of the studs once they're screwed off. Pulling layout from the soft side of the studs is a more accurate method of pulling layout, so many framers establish the soft side of the studs right from the start. The only difference is allowing for half the width of the stud face (usually 5/8 inch) when you establish the layout. To keep things straight, identify the stud layout on the first couple of layout marks at the beginning of the wall, or any time you switch from centers to the soft side.

Step 5: Cutting the Studs to Length The stud layout will tell you approximately how many studs you need per stick of plate. Each stick of plate, unless you've cut it, is 10 feet long. That translates to six studs at 16 inches on center, or five studs at 24 inches on center per stick.

While you're working the top plate, your partner will be cutting the studs to length. As you shoot each stick of plate in place, measure from the floor to the deck at each end, allowing 1/4 inch to 3/8 inch off the hard number (exact length) for the stud length. When the deck's not level, cut only a few studs at a time, using the shortest hard number and allowing only 1/8 inch for play. In extreme conditions, cut the studs individually. Keep in mind that when building with standard (1¼ inch deep) plate as we are here, you'll have 2½ inches of plate to work with. If a stud does come up short, you can lift it up off the floor in the bottom plate to give yourself a little more stud to screw to in the top plate. But don't make a habit of this. Cut your studs accurately in the first place and you won't have to bend the rules like this.

Whether you're cutting your studs by hand or with a chop saw, measure the length of all the studs from the same end. Then stuff the studs with the cut end up every time to keep the stud holes even. Not every wall requires the stud holes line up, but many do. If you make lining up the stud holes a habit, it'll never become an issue for you. I've seen guys get themselves in serious trouble by not lining up the stud holes in a wall that called out for cold rolled channel to run through it. It doesn't take any more time to line up the holes—it just takes paying attention.

Using a chop saw to cut the studs will really increase your productivity. With a new blade on the saw, you can cut the studs by the bundle (for studs up to 6 inches wide). See Figure 5–9. When cutting studs by the bundle, square up the ends of the studs in the bundle at the end you're measuring the stud length from. Use your axe and tap the ends of the studs until they're all even. (And try not to bend them all up doing it.) You'll have to square up the ends of the bundles anytime you're cutting more than one at a time. For example, you could cut two bundles of 1⅝-inch studs at a time.

When using a chop saw, *always* wear safety glasses. I would also recommend that you wear ear protection, but the safety glasses are a *must*.

Step 7: *Stuffing the Studs* Install (or stuff) the metal studs into the wall by tilting them diagonally between the top and bottom plates, then standing them upright in the plate and sliding them in place on layout. Turn the studs in the plate so that the soft (open) side of the studs faces the direction from which you pulled the stud layout. We pulled layout for this wall from the corner, so the first stud is turned hard side facing out to form the outside corner. Your partner will set the bottom of the end stud to the exact corner point and screw it off in the bottom plate. Once set, measure from the 90-degree plumb mark to the hard side of the end stud at

Figure 5–9. Cutting a bundle of 3⅝-inch metal studs at once is a real timesaver. As the chop saw blade wears away (and it will), roll the bundle of studs up into the blade to finish the cut.

the bottom plate. Use this measurement to set the top of the end stud. Clamp the end stud to the top plate and set it to the corner point using the 90-degree plumb mark, then square it in the plate and screw it off. Now stuff the wall studs with the soft side facing the corner.

There are exceptions to this rule. Always turn the slap studs so that the hard side is facing the wall it'll tie into. Door and window studs (the equivalent of the trimmer studs in wood framing) will always be turned with the hard side to the jamb. Here's something to remember. Unlike the trimmer studs, which run from the bottom plate to the bottom of a header, the door and window studs run continuous from the bottom plate to the top plate. We'll look at this in more detail in Chapter 7.

As you stand up the studs in the wall, adjust them to the layout marks and screw them off at both the front and back of the top and bottom plate. The fronts are easy. Just set them to the layout mark. The backs are a different story. You'll have to square them in the plate to maintain the layout from one side of the wall to the other. Square these studs by eye. They don't have to be perfect, but they need to be close (Figure 5–10). If you have trouble eyeballing the studs square in the plate, use your speed square to set them until you develop your eye. It'll be tempting to clamp the studs while screwing them off, and you'll probably see other framers doing this, but don't. Clamping light gauge studs in this situation is a waste of time. Just hold the studs in place with your hand. In Figure 5–11, you can see a stud being gripped tightly in place as it's screwed off. The stud is held tight to the back of the plate by changing your grip slightly. Watch where you put your fingers so you don't run a screw into one of them.

Screwing off the back side of the top plate is a bit of a pain, because you have to reach around to the back side of the wall to do it. In some cases, it's acceptable to run the framing screws from the inside of the stud out through the plate. But make sure you check with your foreman before using this method. The points of the screws sticking out through the plate will cause obvious problems. To make it easier to reach around and run the screw through the plate into the stud, hold your screw gun upside down. Set the back of the screw gun in the palm of your hand, and operate the trigger with your thumb. You'll see this in Figure 5–12.

Stuff as many studs as you can reach from your scaffold, then screw them all off. Your partner will screw off the bottoms as time permits. However, his chief job is to keep you supplied with material and roll the scaffold as you're ready to move. You don't want to waste time climbing up and down the scaffold. Stay put and let your partner move it. On short walls, your partner can often stuff the wall out in front of you as you're screwing off the tops. Once the tops are screwed off, you'll both gang up on the bottoms of the studs. You'll first screw off one side of the wall, then move around to the other side and screw it off, keeping the studs square in the plate.

Step 8: Headering for Obstacles As you work your way down a wall, you're going to encounter obstacles like pipes and ductwork that prevent you from standing up the studs on layout. To overcome these obstacles, cut headers to fit in between the two closest layout studs on each side of the obstacle. We covered cutting typical headers like this back in Chapter 3, so you know how to cut a header correctly. Just let me say that with headers like these, you'll have to cut the shoes squarely.

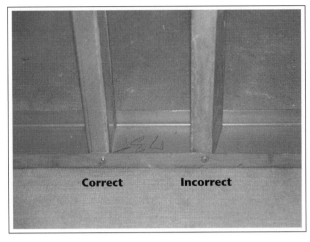

Figure 5–10. Two studs, one correctly and one incorrectly screwed off in the bottom plate. The stud on the right is too crooked.

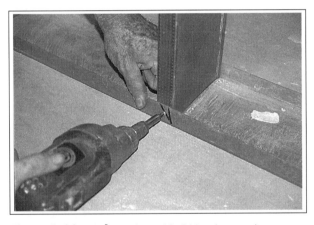

Figure 5–11. A 3⁵⁄₈-inch stud held in place on layout as the stud is screwed off in the bottom plate.

Figure 5–12. You can screw off the back side of the stud by flipping the screw gun upside down and operating the trigger with your thumb. Hold the stud in place by hand.

The length of the headers is given as "in between." For example, a measurement of 30³/₄ inches tells you the distance *between* the two shoes. The shoes are always approximately 3 inches long. Whenever you're cutting a header to fit between two layout studs, the measurement should always be some number that ends in ³/₄ inch. That's because the "in between" length will be 1¹/₄ inches shorter than the on-center layout measurement of the two studs you're spanning. If you're working with a 16-inch on-center layout and you're headering for one stud, the header would be 30³/₄ inches in between. If you're headering for two studs, the header would be 46³/₄ inches in between, and so on. The 1¹/₄ inches is the width of the face of the studs, and that's what you're subtracting for. It works the same on a 24-inch on-center layout.

Any time the wall you're building will *top out* (rock to the deck), you must frame in a header and cripples above the obstacle as well. You'll see some framers who just leave out the top header and cripples. They'll tell you that if the rockers need a header, they can put it in themselves. Make your own call. Any wall should be headered above any obstacle, but some framers cut corners by leaving out the top header and cripples. If you decide to leave out the top header, make sure your foreman approves of the idea. You don't want to have to go back to an area a second time to finish that wall.

To put the header in the wall, just tilt it diagonally and slip the shoe onto the studs as you tilt it into the level position. Now slide the header into place, leaving approximately a 1-inch gap between the header and the obstacle. Level it by eye. Screw the header off on both sides of each shoe. The 1-inch gap is especially important on ductwork. Any time ductwork penetrates a fire-rated wall (fire wall), the tinners will add a damper to their duct where it penetrates the wall. After the wall is rocked (also leaving a gap of ³/₄ inch), the tinners will add a metal collar to the duct. If the collar doesn't slide easily into place, they'll make it fit using their fine adjustment tool — a 2-pound sledgehammer.

Have your partner cut the cripples to length, allowing ³/₈ inch, as you lay out the header. Pull layout across the header from the soft side of the stud the header is screwed to. Work from the same direction the rest of the layout was pulled from. Stuff your cripples in the wall and get them screwed off on layout, and the header's a done deal.

When framing in a header around an obstacle, you'll often have a layout stud hit just under the obstacle. This leaves a large void beside the obstacle, giving the drywallers nothing to screw the rock into. When this situation comes up, cut a piece of stud or plate with ears on each end to fit in between the headers. Screw it off, leaving a 1-inch gap off the obstacle, as you can see in Figure 5–13.

On extremely wide ductwork, add extra studs off layout at one or both sides of the duct, just so you'll have something to header to. A header much over 6 feet long will leave a weak section in your wall. Additional studs off layout at the sides of the duct help to keep the wall solid. Be sure to maintain the 1-inch gap off the duct, measuring off the closest layout stud at the top and bottom plate to plumb the extra studs. Measure the header length at the top or bottom plate, measuring in between the added studs the header will be screwed to.

Once your header is screwed off to the studs, establish the layout across the header. To do this, measure from an edge of the stud you added to the closest layout mark at the bottom plate. Let's say it's 10 inches. Now, burn 10 inches past the same edge of the stud and mark the header at 16 inches, then pull layout for the rest of the cripples from this point.

Figure 5–13. Two layout studs hit under this heater unit. The header over the unit leaves a large gap in the framing at each stud. A stud with ears has been added on the left of the heater, and another will be added to the other side.

Step 9: Ending the Wall There are a few different ways to end a wall. The most common is an inside or an outside corner. We started this wall with an outside corner, and we'll end it with the inside corner of an intersecting wall. We stopped the plate $3/4$ inch short of the intersecting wall line to allow the drywall to slide past. We slid the slap stud into the wall with the hard side of the stud facing the intersecting wall, and set it tight to the end of the plate. Finally, we ran one screw in the slap stud at the top plate, just to keep the stud from falling out of the wall. Figure 5–14 shows the wall ended at an outside corner with an intersecting wall.

Freestanding Walls

The term *freestanding* pretty well describes this style of wall. These walls are generally built with studs just long enough to get the top of the wall, about 6 inches above ceiling height, instead of running to the deck. In most cases this eliminates the need to cut the studs to length. Because these walls don't run to the deck, you add stability to the wall with kickers running from the top plate of the wall to the deck or bar joist. The kickers will run on approximately a 45-degree angle, spaced about 4 feet apart by eye.

You lay out and shoot down the bottom plate for a freestanding wall just like the bottom plate for a wall to the deck. So we'll skip plating the floor and begin this discussion with the bottom plate shot down and the stud layout pulled. Let's also assume that a grid ceiling will be installed later with a height of 8 feet. So we'll build our walls out of $3^{5}/8$-inch wide studs. We'll use a 9-foot length to get us well over ceiling height.

Step 1: Setting the Ends To begin the wall, set the end stud in the bottom plate and slide it in place. Plumb the end stud with your level, tie it in to the already-framed wall, or if it's a masonry wall, shoot it to the wall with a shotgun. Space the pins about 24 inches apart, shooting the pins right into the mortar joints of the masonry wall. Anytime you're shooting to masonry walls, even brick, you'll shoot to the mortar joints. The force of the shotgun will just blow through a cinder block, and bricks will often split or shatter.

As one partner is shooting the end stud in place, the second will scatter bundles of studs out along the bottom plate. Set the bundles about 4 feet off of the wall to leave easy access to the bottoms of the studs. Before you get up on your scaffold, check to make sure you've got plenty of loads, pins and framing screws.

Step 2: First Stick of Top Plate To get going, climb up on your scaffold and slide the first stick of top plate onto the end stud, as your partner stands up a stud at the opposite end of the plate. See Figure 5–15. Pick the top plate up off of the end stud about $1/4$ inch and screw off both the front and back of the stud. Then balance the wall while your partner sets the stud he stood up on layout, and screws off both the front and back of the stud. With the front and back of the stud screwed to the bottom plate, the stud will stand by itself. Now pick the plate up off the top of the studs about $1/4$ inch and clamp, then screw, the top plate to the stud.

As your partner holds the stud steady and as plumb as he can get it by eye, measure from the top plate up to the deck or bar joist to get the

Figure 5–14. The wall ended with a slider that'll later be tied into the intersecting wall.

kicker length. Remember, the kicker must run on about a 45-degree angle for maximum strength. You'll usually shoot the kickers to the same point on the deck or bar joist, so the kicker lengths are the same down each wall.

Remember, the kickers are spaced approximately 4 feet apart. Clamp the flattened shoe of the kicker to the top plate and you can balance the wall as you're shooting the kicker to the deck. That frees your partner to cut the next kicker and stuff up the wall. Once you get the kickers at 4 and 8 feet shot to the deck, tack them to the top plate with a single tek screw (Figure 5–16).

Step 3: Pulling Layout   With your first section of wall stuffed up and kicked off, your next move is to establish the layout along the top plate. Plumb up the layout with a plumb bob, or stand a stud up outside the wall with the hard side of the stud up against the top and bottom plate. Set an edge of the stud to a layout mark on the bottom plate, and use a level to plumb up the stud and layout, as shown in Figure 5–17.

Once you've established the layout at the top plate, set the first stud on layout and screw it off at both the front and back of the plate. Then pull layout from the soft side of this stud. Finally, set the rest of the studs on layout and screw them off at both the front and back to the plate. Your partner will get your next stick of plate and kickers ready as you're working the top of the wall, then screw off the bottoms of the studs as time permits.

Step 4: Splicing Joints in the Plate   As soon as your partner rolls the scaffold down the wall into position, splice your next piece of plate onto the last one. Your partner will again stand a stud in the bottom plate on layout near the end of the top plate to support that end.

Clamp the two sticks of plate together at the joint, and run a single screw into the joint to tack the plate together. Now shoot your kickers to the deck, maintaining the 4-foot spacing. As soon as you get the first kicker shot up, tack it to the top plate to balance the wall. Then your partner will stuff up the wall and screw off the bottoms of the studs as you're working the tops. Pull layout, again pulling off the soft side of the

Figure 5–15. With the end stud of this freestanding wall plumbed and tied in, the first stick of top plate is set on top of the end stud and first layout stud. Setting the first layout stud and screwing it off on both sides of the bottom plate will keep it standing up as you pull layout along the top plate.

Figure 5–16. This freestanding wall is held in place by kickers spaced 4 feet apart. A diagonal brace is also used to prevent the wall from racking. The brace will be removed as the wall is rocked.

Figure 5–17. Here I'm using one of the wall studs as an extension of my 4-foot level to plumb the stud layout from the bottom plate up to the top plate of this freestanding wall.

last layout stud screwed off in the top plate. Get the studs set on layout and screwed off.

What if you run into ductwork or other obstacles that interfere with a layout stud? Simply header around it, just as you'd do with any other wall. But that's another advantage of this type of wall. Most of the other trade's materials will go *over* the wall, not through it.

Step 5: Framing a Corner Now you're ready to frame the corner. The first thing you'll do is figure your plate length. The quickest way to get the measurement is to use a stud and a 4-foot level, just like plumbing up the layout. Have your partner hand you up the stick of plate with the splice cut in one end, and splice it to the last stick of plate. Be sure to shove the splice tightly together. Your partner will stand a stud up outside of the plate with one edge lined up on the intersecting wall line, and plumb the stud with a 4-foot level. Once he's got it plumb, mark the top plate at the same edge of the stud. You've got your length. Remember, the plate doesn't need to be cut exactly to the corner point. So cut the plate about 1/4 inch short of your mark to give yourself a little play to work with.

Resplice the joint in the top plate and clamp it off for now, after you've jammed it tightly together. Your partner will stand up the end stud in the bottom plate, to hold up one end of the top plate. Now line up the kicker for the end of the wall, with the shoe of the kicker hitting the top plate about 4 inches back from the end. Shoot the kicker to the deck. Clamp the shoe of the kicker to the top plate, and your partner is once again freed up to stuff up the wall. Now get the rest of the kickers shot up to maintain the 4-foot spacing, then tack the kickers and joint in place. Pull layout and get the studs screwed off, and you're ready to set the corner.

With everything screwed off, plumb the corner either by shooting up a laser beam or using a 4-foot level. Use the end kicker to make the needed adjustments for plumbing the corner. If you keep the clamp in place and use light pressure to hold the kicker and plate together, you'll be able to make very fine and accurate adjustments. Once you have the corner plumbed, you're ready to permanently screw off the kicker, forming a triangle with the framing screws. Run two of the screws in right at the fold

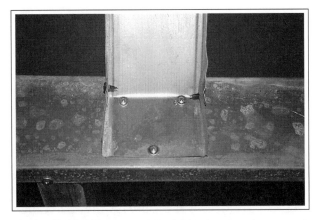

Figure 5–18. The shoe of this kicker is correctly screwed down to the top plate.

of the shoe, and the third out at the end of the shoe (Figure 5–18). With the corner plumbed and kicked off, you're ready to pull a dryline and set this wall.

Step 6: Kicking Off the Wall To complete the wall, you'll need to pull a dryline down the wall, and straighten the top plate. In Chapter 5 we covered setting up and using a dryline. Once you have the dryline set up, work your way down the wall, setting the stud closest to each kicker to the dryline. That's what you see in Figure 5–19. You'll also add three more framing screws to the joints in the plate as you go, locking the joints in place. First get the wall set and the kickers screwed off on both sides of each joint, then run four screws down through the top of the joint. Put one screw in each of the four corners of the joint. Even though you're setting the wall to a dryline, have your partner sight down the wall to double-check the joints in the top plate.

Step 7: Turn the Corner Now you're ready to turn the corner. Your first stick of top plate will form the corner itself, so your partner will cut back the inside leg of the plate and fold up the leg to let the plate overlap at the corner (Figure 5–20). If the leg that's cut and folded looks like it might interfere with the shoe of the kicker, either fold it up out of the way or cut it off.

Set the corner in place, overlapping the last stick of top plate of the intersecting wall, as your partner stands a stud in the bottom plate on layout near the opposite end and screws it off. Slide the corner end of the plate tight to the corner stud, with the very end of the plate about 1/4 inch shy of the actual corner. Run a single tek screw through the leg of the plate into the corner stud. You can also see that in Figure 5-20.

Figure 5-19. Pat is using stilts to work this freestanding wall. In the background you can see the kickers that run down from the deck, bracing the wall. The brace Pat is clamping is a flat horizontal kicker run over from the wall across the hall.

Figure 5-20. The completed outside corner.

If the deck above is a flat, even concrete deck, your partner will already know the kicker length and have at least one kicker cut and ready for you. Clamp the kicker to the top plate about 8 feet from the corner. Eyeball the wall plumb and shoot the kicker to the deck. You're off to the races! The corner will remain tacked just like the joints until you set the wall to the dryline. Follow the same process and work your way to the next corner. This process is very much a *system*. It's a prime example of working smarter, not harder. Each partner must anticipate the other's next move. The top partner knows when the ground partner needs a measurement, and the ground partner knows when the top partner needs a piece of plate, and so on.

Step 8: Jambs As with any wall, freestanding walls will include door and window jambs that you'll have to frame in. Because we'll cover hollow metal jambs in detail in Chapter 6, I won't go into setting the jambs here. What we do need to look at is how a jamb will affect your wall.

Typically, a window jamb won't affect a wall unless it's extremely large. Door jambs, on the other hand, will require additional kickers right above the door and king studs. You'll usually need an extra kicker on both sides of the jamb (Figure 5-21). This is another spot where some framers take a shortcut and only kick off one side of the jamb. If you're going to use only one kicker, you'll want the kicker on the latch side of the jamb. In either case, the 4-foot spacing for

Figure 5-21. The door jamb in this freestanding wall is kicked off on both sides of the jamb. Notice that the jamb has a satellite window.

the kickers will resume 4 feet from the door jamb kicker.

Following the procedures outlined in each of these steps, continue working your way on around the wall, one stick of top plate at a time, to the far end stud. Figure 5–22 shows the completed wall.

Chase Walls

In commercial buildings, the general public rest rooms for both men and women are usually built separate from the surrounding walls. You'll frame the rest rooms as one large room that's separated by a chase wall. A chase wall is two walls built close together, leaving a small void in between them. In the case of the bathrooms, the void (or chase) between the walls is where all the plumbing lines for the toilets are housed. You can bet that they pack as much as possible into as small a void as possible. That's why we'll focus this discussion on chase walls that divide rest rooms. They'll be as challenging as any chase walls you'll ever have to build.

The chase walls are framed as freestanding walls unless the wall requires a fire, smoke or sound rating. Then they're framed as walls to the deck. We've already covered both the freestanding wall and walls to the deck in detail, so you know most of what you need to know about a chase wall.

When the plans call for the chase walls to be framed to the deck, it's often acceptable to frame only one of the two chase walls to the deck. Your foreman will make the call. Then you'll rock the top of the shorter wall over to the back of the tall wall, which caps the void and continues the rating of the wall. Run a horizontal control stud across the back side of the tall wall at the same height as the top plate of the short wall. This stud will not only carry the drywall shelf, it'll also support the rock for the back side of the wall until it's screwed off.

Be systematic about framing the rest rooms. You'll want to frame the perimeter walls of the rest rooms first, then build the chase walls to tie into them. To save yourself some time tearing down and rebuilding the scaffold, you can frame only the side walls of the rest rooms that the chase wall will tie into. Or frame the entire perimeter and don't stuff up the end walls. This will allow you to roll the scaffold in and out without tearing it down (Figure 5–23). Complete the end walls from outside of the rest rooms after you've built the chase walls.

Chase walls are quite a challenge. But just knowing what you're in for and what it takes to get the job done makes it a lot easier to get through. So let's focus on some of the oddball techniques it often takes to frame a chase wall. The bottom plate is where you'll run into problems. Once the wall lines for the chase walls are snapped, the plumbers will often set their rack for drains and water lines in place on the wall line and anchor it. The rack is a pretty overwhelming obstacle. In fact, it's common for the

Figure 5–22. A completed freestanding wall.

Figure 5–23. This is a simple chase wall framed up to a fire lid. The front wall of the rest rooms was left unframed to allow access to the chase walls.

bottom plate to be pieced in and even headered, as you can see in Figure 5–24. At times you'll even have to add studs off layout, just to have something to header to. You'll also have to notch both studs and plate to get around some of the obstacles.

The way the walls are rocked makes a big difference on the amount of trouble you need to go through. If the drywall has to be installed vertically (which is usually the case), you'll have to install a stud on layout at least every 48 inches on center. Set the layout studs in between (at 16 and 32 inches) wherever you can get them, as long as they're somewhere in the vicinity of layout. Horizontal installation of the drywall greatly reduces the number of studs that have to hit the actual layout. Once you get a layout established, the drywall can break on the same studs on both sides of the walls.

Be sure you let your foreman know how you're going to frame the chase. There may be other complications that affect the way you frame the wall. An angry or very meticulous framing inspector will make a big difference in how you frame a wall, and what you can get away with. The amount of extra work involved in framing the chase may also require an extra work order. This is something your foreman must deal with *before* you start the work.

Figure 5–24. Here's a tougher chase wall. Several of the layout studs are headered for close to the bottom plate.

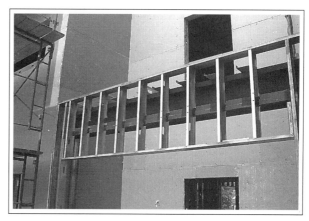

Figure 5–25. The first side of the framing is completed. The weight of this arch will be carried by the lookouts shot to the top of the red iron. If you look closely you can see that the studs and cripples had to be notched for the steel.

Arches

Arches are a common part of metal stud framing. We'll work through framing an arch in a structural stud wall. As we work through the job, I'll point out the few differences in framing an arch with light gauge metal studs. The process for framing arches, whether they're large or small, structural or light gauge, is pretty much the same. Frame most or all of the wall around the arch, and hang the drywall. Then cut out the arch in the drywall and finish framing the arch to the rock. Now let's go through the steps.

Step 1: Frame Up Both Ends and Above the Arch You'll start framing the arch after the walls on both sides of the arch are framed up. Then span the walls on each side with a top plate and fill in any needed loadbearing framing above the top of the arch, as shown in Figure 5–25.

On a light gauge wall, run the cripples above the arch from the top plate down to the radius plate of the arch. If the light gauge wall is a freestanding or a deflection wall, you'll need to frame in a box beam header above the top of the arch to carry the weight of the framing and drywall.

Step 2: Run the Gyp Board Wild Next, get the exterior of the wall rocked. On most commercial projects, they'll have crews of rockers to hang the exterior gyp board. You'll get all the arches framed to this point before they rock the entire exterior. Then they can run the gyp board wild over the archway, covering it completely (Figure 5–26).

On a light gauge interior wall, once the arch and king studs are in place with the bottom header, and cripples framed in, rock one side of the wall with two or three sheets to cover up the opening for the arch. Make sure the sheets you hang break on the appropriate layout studs, so the wall can be completed without a hassle.

Step 3: Cut Out the Arch With the exterior gyp board hung, establish the center of the arch. The center of the arch is the same as the center of the rough opening of the arch. Remember to allow for any 2-by material that it's wrapped in. Using your level, plumb the vertical center up the wall well past the center or pivot point of the radius.

Once you've got the vertical center line drawn out, establish the horizontal center or pivot point for the arch. First establish the top of the arch, then measure back down the horizontal center line a distance equal to the size of the radius. Once you've found the center point, run a drywall screw into it. Now hook the end of your tape to the screw and strike the arc (Figure 5–27).

In most cases, if an exterior wall has any arched windows, all of the windows will be arched. Then you just establish a common elevation for the top point of all the radiuses using a water level or laser. This will maintain a constant elevation for the arches and tie all the windows together.

With the radius of the arch drawn out, cut it out with a router (Figure 5–28). As you cut around the radius, you'll feel the router bit come in contact with the window studs which establish the sides of the window. The bottom header in exterior window rough openings will commonly get doubled up with 2-by material. When the sides of the rough opening also require 2-by material, plumb a line down from the sides of the radius, then cut along the line. Leave the rock running wild. The 2-by material usually isn't added until much later.

Step 4: Plate the Arch With the arch cut out, it's time to plate the radius. To get started, use a chop saw and cut some radius plate out of standard structural or light gauge plate. Space the

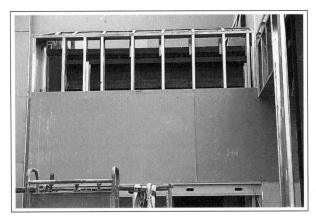

Figure 5–26. Notice the leveled stud screwed off to the wall studs, spanning the archway opening. The drywall will be screwed to the framing on each side and the top. The horizontal stud carries the weight of the drywall that floats in the middle, as well as keeping everything level.

Figure 5–27. Hank is drawing out an arch with his tape and pencil. The end of his tape is hooked on a screw head run into the rock at the center point of the radius.

Figure 5–28. This is a good example of an arched opening cut out in the gyp sheathing. The framing sticking down into the arch is an optical illusion. The arch is 24 inches wide and the framing is inside the top of the arch.

cuts in the legs of the plate about 3 or 4 inches apart, and make sure they're square across the plate. Now cut the pieces of radius plate long enough to form the radius, plus a couple inches on each end to tie into the window stud. As you cut the plate, keep it tight against the fence of the saw to keep the cuts square. To cut just the legs of the plate, you'll roll the plate slightly up off the saw table. Just make your cuts slowly and you won't have any problems.

With the plate (or in this case utility angle) cut, you're ready to form the radius. This will take one partner on the inside and the other on the outside of the wall. Slide the utility angle in place around the radius of the arch. As the ends of the utility angle slip over the window studs, you'll often need to back out a couple of the drywall screws to allow the leg of the angle or plate to slide in behind the drywall. The partner inside will line up the angle, exactly following the radius, and hold it tightly against the back side of the gyp board, using the head of their hatchet. The partner outside will be screwing off the angle, spacing the screws 3 or 4 inches apart (Figure 5–29).

Figure 5–29. The utility angle is screwed off every 3 to 4 inches, holding it to the radius cut out in the gyp sheathing.

Step 5: Complete the Framing Now you've got the utility angle formed to the radius of the arch and screwed off on both sides. It's time to complete the framing by spanning the utility angle with drywall channel. Cut the DWC about $1/4$ inch shy to allow a little play. The layout for the DWC will depend on how tight the radius of the arch is. In Figure 5–30, the DWC was spaced 8 inches on center to form a smooth radius.

With the framework of the arch completed, finish up by rocking the belly of the arch. Start on one side and work your way up, over and down the radius of the arch. As you tack up the ends of the rips, slowly push the middle of the rip up into the curve of the arch. On really tight radiuses, wet the drywall to make it a little more pliable. Screw off the rips, again spacing the screws 3 or 4 inches apart along the outside edges and one in the middle of each stick of DWC. This will form a nice smooth arch (Figure 5–31).

Before we wrap up this discussion, let's take a quick look at one of the trade's latest advances: the *radius plate bender.* The radius plate bender crimps the plate or studs, bending the framing members to form a radius. You adjust the spacing

Figure 5–30. The utility angle is screwed off on each side of the arch and spanned with DWC spaced 8 inches on center. That completes the framework of the arch.

Figure 5–31. The arch is completed and ready for the EIFS (Exterior Insulation and Finish System) crew.

Figure 5-32. Here's a very complex radius drop framed using a radius plate bender instead of patterns and radius plate.

and depth of the crimps to form different sizes of radius. The instructions that come with the tool explain it all in detail. The radius drop in Figure 5-32 was formed with a radius plate bender. It saved countless manhours of cutting patterns and radius plate, then screwing them together.

Chapter 6

Hollow Metal Jambs

In this chapter we'll take a look at the hollow metal jambs used in commercial metal stud framing. The responsibility for setting the jambs will vary from project to project, depending on the contract for each job. The jambs may be set by the General Contractor on the project, or the metal stud framing outfit. In most cases, the framers set the jambs. In either case, to be a metal stud framer you have to be able to set the jambs properly. In this chapter we'll cover the basic methods, and a few variations.

The hollow metal jambs (which are fire rated) eliminate a lot of the time normally spent to frame and trim out a wood jamb. The metal jambs come complete, trim and all, in one package. There are two basic types of door jambs: the typical one-piece jambs that you set as, or before, the wall is framed, and the three-piece jambs that you set after the wall is both framed and rocked. You'll set all the window jambs as you frame the walls.

All one-piece hollow metal door and window jambs are fastened to the framing with clips. Since the clips are probably the most important component in a hollow metal jamb, that'll be a good place to begin this discussion.

Clips

The clips are the bridge between the jamb and the framing. They're available in three different styles. Figure 6-1 shows:

- The standard clip is either welded in the jambs at the factory, or snapped tightly into the jambs on the job site. The broader section of the clip snaps in place in the jamb, and the narrow side of the clip screws to the door and window studs with S-12s.

- The clip with straps attaches to the jamb, just like the first. The straps (originally designed for use on wood stud framing) wrap around both the door and king studs. The installer screws both straps to the door and king stud, as in Figure 6-2. Keep in mind that it's common simply to cut the straps off and use them as standard clips.

- The adjustable clip can be set for different widths of jambs. You set this clip to width and snap it together, then install it into the jambs just like the first two.

Because the standard clips are much more common, we'll focus on them.

Field Installation

When the clips aren't installed at the factory, you simply slide them into the jamb diagonally, then turn them square, which locks them in place in the jamb. See Figure 6-3. The clips must fit tightly in the jamb. If a clip is a little loose, bend out the tabs on the ends of the clip with your lineman's pliers. These clips come in several widths that fit the different widths of door and window jambs.

For door jambs, you'll install three clips per side. Install the first clip about 12 inches up off of the bottom of the jamb, then one in the middle. Put the last clip about 12 inches down from the top of the jamb. The top (or head) of the jamb doesn't need any clips.

On window jambs, the size of the jamb determines the number of clips you'll need. Just maintain an approximate 2-foot spacing between the clips. In most situations, two clips on each vertical side of the jamb and two on the bottom will do the job. Twist the clips into the window jambs, just as in the door jambs.

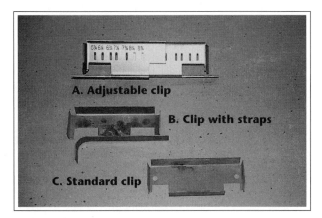

Figure 6-1. These are the three most common types of door clips.

Figure 6-2. The strap on the door clip has been folded around the door and king studs and secured with a 34 through the strap into the door stud. A second 34 will run through the strap into the king stud after it's shoved tight to the door stud. We'll do this to all six door clips, on both sides of the studs.

Sooner or later, any metal stud framer will run into lead-lined walls. They're common in hospitals, usually built around examination and X-ray rooms. This lead lining affects not only the walls (which are rocked with lead-board), but the jambs as well. Any jambs that are part of the lead-lined walls are also lined with a thin layer of lead. That's usually done by the manufacturer. If you're lucky, they'll also

Chapter 6: Hollow Metal Jambs   97

put in the clips. If not, you install the clips as usual, but it's a tight fit because of the lining. The biggest concern you'll have is getting the clips in without tearing or piercing the lead. The clips will bend a little under the stress, allowing them to turn into place.

Door Jambs

As we prepare to set a typical jamb, let's assume a couple of things. First, let's say we're setting a standard 3-0 jamb (that's 3 feet, and it's actually the door size) in a 25 gauge $3^{5}/_{8}$-inch wall. Second, let's assume we're setting the jamb as we're framing the wall, which is usually the case. First you'll shoot down the bottom plate, leaving a 40-inch gap in the plate. The gap in the plate is always 4 inches wider than the jamb size (which is the door size, or the *inside* width of the jamb). The outside width of the jamb is 2 inches wider on each side, or 4 inches total. Figure the header length the same way.

Figure 6–3. The door clip in the middle of this door jamb is turned diagonally as it's installed in the jamb. The clip at the top of the jamb has been turned horizontal to the jamb, locking it in place.

Step 1: Choosing the Correct Jamb   All doors are numbered in the prints, and each jamb comes from the manufacturer with a tag just below the bottom hinge slot. The number on the tag should correspond with one of the numbers in the prints. The tag will also have the proper swing marked on it — either R.H. or L.H. If you'll recall from the chapter on laying out the walls, the door number and swing should be written on the floor through the opening on the bottom plate.

Step 2: Set the Jamb and Shoot It Down   Set the jamb in place in the door opening. Then use your level at the top of the jamb to ensure that it sits level on the floor. A 2-foot magnetic level really comes in handy here. That's what the framer's using in Figure 6–4. The magnet frees up your hands and you can read the level vial through a window in the level's frame from under the level. If the jamb is sitting level, you're ready to go. If it's not, shim the low side of the jamb with cardboard shims or pieces of duct strap until it's level (Figure 6–5). If a jamb needs to be shimmed up more than $1/4$ inch, let your foreman know. In most cases the jamb can only be shimmed up so far.

Figure 6–4. This hollow metal jamb is being set with a 2-foot magnetic level.

Once the jamb is level, it's time to shoot it down. As your partner balances the jamb on both sides, adjust the bottom so there's an equal gap between the plate and the inside lip of the jamb. You can see that in Figure 6-5. The gap should be equal to the thickness and number of layers of the drywall that'll be hung on the walls later. With the gap set on the first side of the jamb, shoot down the door jamb cleat with a single pin for now. You'll usually use a shotgun with yellow loads and concrete pins to shoot down the jambs. Occasionally, however, you'll use a hammer drill (vibrating drill) and concrete screws to screw the cleats down. Now, move to the other side of the jamb, set it and shoot it down with two pins. Then go back and add a second pin to the first side of the jamb.

Step 3: Door and King Studs The door and window studs are the same as trimmer studs in wood framing, except that they run from floor to deck. Keep in mind that in certain structural framing situations, metal stud trimmers are used. First, get your door and king studs cut to length. Because the wall in our example is framed with 25 gauge studs, use 20 gauge studs for the door and king studs to stiffen up the jamb and prevent the wall from rattling when the door shuts. Slide the door studs with the hard side to the clips on each side of the jamb to steady it. Next set the gap between the door stud and the lip of the jamb. Clamp the door stud to the door clip, centering the door stud in the jamb (Figure 6-6). As soon as you get the gap set, have your partner tack it, running a 34 (another trade term for a self-drilling framing screw) through the stud into the clip. Then run two screws into each clip. Set both sides of the jamb and get all six clips screwed off. Slide the king studs in place with the hard side tight to the door studs on each side, and screw off all four studs in the bottom plate.

You'll need to be on the scaffold to complete the jamb. First, plumb the jamb with a 4- or 6-foot level, and screw off the door studs in the top plate. The jamb doesn't have to be exactly plumb, but it should be close. The drywallers will plumb the jamb as they rock the walls. Slide the king studs tight up to the door studs and screw them off. Now you're ready for the header — which the partner on the ground should have cut and ready for you.

Figure 6-5. The cleats of this door jamb have been shimmed with pieces of duct strap. Notice that the door jamb is centered on the bottom plate.

Figure 6-6. Here the door stud is clamped to the door clip, and it's being centered in the jamb. Instead of measuring the gap, you can simply shove scrap drywall in each side of the jamb.

You have to cut the shoes of the header square, or the header will twist on the door studs that you fasten it to. Remember that the in-between length of the header will be 4 inches longer than the width of the inside of the jamb. A 3-0 (or 3-foot) door will require a header length of 40 inches "in between." Tilt the header diagonally between the door studs, then as you tilt it back level, slip the shoes on to the door studs. The header needs to be even across the top of the jamb.

To set it, lay your two pieces of scrap drywall across the top of the jamb and slide the header down on to the rock, as shown in Figure 2-4 back in Chapter 2. Do this on each side. Give your partner the length of the cripples, then screw off the header to the door studs with one screw on each side of each shoe.

To establish the layout across the header, have your partner measure from the hard side of a door stud to the center of the closest layout stud at the bottom plate. Now, burn that number past the hard side of the same door stud at the top of the jamb and mark the layout across the header. When the rockers plumb the jamb, the layout across the header will also be plumb automatically. Install the cripples on layout and get them screwed off as your partner cuts the spreaders off the bottom of the jamb with a chisel. And this jamb's a done deal. Figure 6–7 shows the completed jamb.

Speaking of spreaders, there's one complication you might face: The factory spreaders don't always last until you get the jambs set. When the spreaders pop loose before you set the jamb (which is common), cut yourself a plywood template/spreader and use it to set the jamb, as seen in Figure 6–8. Make sure you cut the ends of the plywood spreader square.

Headered Door Jambs

Throughout this section, I've mentioned that the door studs will run continuously from bottom to top plate. You'll inevitably encounter jambs that will require the door and king studs to run to a header because of obstacles in the way. See Figure 6–9. This is a less-than-ideal situation, but one that you'll come across occasionally. You may also run into situations where the door and king studs on both sides of

Figure 6–7. Here's a typical completed door jamb. Notice that the top plate of this wall is deep leg and the door and king studs are left unscrewed. On a nondeflection wall the door and king studs are held together tightly as they're screwed off.

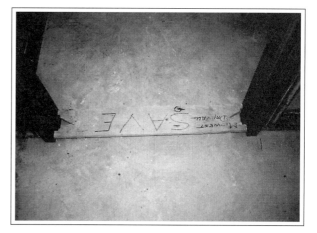

Figure 6–8. This is a plywood spreader being used to set a hollow metal door jamb after the factory spreader popped out.

Figure 6–9. The door and king studs can't run to the top plate because of the sprinkler pipe running right above them. We ran a header between the door and closest layout stud, and the door and king studs run up to the header.

the jamb run to a header. In a case like this, just run your header over the entire door framing, then frame up the jamb as you normally would. Once the jamb is framed in, run kickers down to the door and king studs above ceiling height to stiffen up the jamb.

Door Jambs on Computer Floors

In the commercial trade, there's no doubt that eventually you'll frame up a job that has a computer floor. A computer floor is a tile flooring system that's elevated a couple feet above the concrete floor. Cables run under it, so they're easily accessible for repairs. Any area where there will be a lot of computers and other electronics usually needs a computer floor installed. The prints will specifically call out the computer floor if one is required.

The computer floor is as close as you'll get to a perfectly flat and level floor. It goes in after the walls are framed and rocked. This means you won't be shooting the door jambs down to the concrete floor. But that's only one of the differences between this and a normal job. Let's take a look at the rest.

Step 1: The Framing The first difference you'll notice is that the bottom plate will run continuously under the door openings, and the jambs are headered underneath (below the floor tile) after they're set. The wall will frame up just like any other. As you stuff the studs, go ahead and also install the door studs and get them screwed off in the bottom plate.

To make sure the bottom of the jamb and the computer floor have a common elevation, it's important to set the jambs to a benchmark or laser. Once the laser is set up, you and your foreman should work with the foreman who's installing the floor, or the General Contractor's superintendent. They've got to establish the floor elevation, so that you can in turn establish the elevation for the bottom of the jamb. It should be no more than $1/8$ inch above the computer floor. Then, with the door studs fairly plumb, use the benchmark or laser to establish the jamb elevation and mark it on the door studs. Because the jamb doesn't rest on the concrete slab or the computer floor, you can't shoot or screw down the bottom cleats. That leaves the bottom of the jamb loose. That means you should add a fourth clip down at the bottom on each side of the jamb.

Step 2: Setting the Jamb Spread the door studs enough to set the jamb in between them, and slide the jamb in place. Get the jamb into the approximate position and elevation, then clamp it in place by clamping the clips to the door studs. Now adjust one side of the jamb to the exact elevation and center it up on the door stud at both top and bottom. Once set, screw off the door stud on this side of the jamb with two S-12s per clip.

Now that you've got one side of the jamb set and screwed off, level the head (top) of the jamb with your level. Clamp the clips to the door stud on the opposite side of the jamb. Make all your needed adjustments to set this side of the jamb, get it screwed off, and you're ready to header in the jamb.

Step 3: Header In the Jamb The top of the jamb will header in just like any other jamb. The difference is that in this case, you'll add another header below the jamb. Cut your bottom header to length and slip it onto the door studs under the jamb, then slide it snug against the bottom of the jamb. Then cut the cripples and fasten them in place on layout. See Figure 6–10.

To complete the jamb, add the header and cripples above the jamb, as well as the king studs. Because this door is basically suspended, you can't shoot the cleats down to the floor. Screwing off the extra clips you added to the bottom of the jamb on each side will do the same job. To cut the spreaders off the bottom of a jamb, it's best to use a reciprocating saw or a hacksaw. Trying to cut off the spreaders with a chisel will just bend up the bottom of the jamb.

Tight-Fitting Jambs

It's inevitable that you'll run into a door jamb that will sit right next to a concrete wall or a vertical member of the red iron superstructure. This usually won't leave you with enough room to screw the door stud to the clips. To overcome this problem, you'll need to screw the door stud to the clips before you shoot down the jamb. And that brings up another problem. If you screw the stud in place, how are you going to get the shotgun in position to shoot down the cleat? The answer is simple. Just cut the door stud to run only from the top plate down to the bottom door clip, leaving a void in the door stud from the bottom door clip to the bottom plate (Figure 6–11).

Set the jamb in place, and check for level. You won't need to shim the jamb or level it right now, but if the jamb is out of level it'll affect your measurement. Now measure from the top plate down to the bottom of the bottom clip on the tight side of the jamb.

Figure 6–10. A box beam header is used to carry this door jamb.

Figure 6–11. The door stud is cut short and screwed off to the door clips, with the door jamb set in place ready to be shot down.

Step 1: Set the Stud Set the door stud (for the tight side of the jamb) in place against the clips and allow the stud to slide down about $3/8$ inch past the bottom clip. Remember, you cut the stud tight from the top plate to the bottom of the bottom clip. That $3/8$ inch will allow for any adjustment you'll have to make to the jamb to make sure it's level. Now adjust the gap and clamp the stud to the clips, then screw the stud off to the clips using two S-12s per clip. See Figure 6–12.

Once the door stud is screwed to the clips, measure the void between the end of the door stud and the bottom of the jamb. Cut a piece of plate about 12 inches longer than that and slip it onto the stud above the bottom clip. After the jamb is set, slide this piece of plate down the stud into the bottom plate to fill the void.

Step 2: Shoot It Down In many cases where you're having to go to all this trouble, the king stud on the side you're working will be shot to the masonry wall. Sometimes this will leave a little space between the king and door studs, making it a bit easier to get the shotgun and your hand behind the jamb. The important thing here, though, is that if the king stud must be shot in place, do it before you set the jamb. In this situation it's common to also cut off the king stud, just like the door stud, to get your shotgun into position.

Now set the jamb in place, level it, then adjust the gap between the bottom plate and the inside lip of the jamb. With the gun ready to fire, slide it into position behind the jamb with the barrel on the cleat, as shown in Figure 6–13. You really have to be careful here. Make sure you've got a good firm grip and enough pressure on the gun to prevent it from kicking your hand up into the stud. Like all cleats, you need two pins in it — one on each side of the cleat.

To finish this side, slide the piece of plate you added to the stud earlier down the stud and into the bottom plate. Screw off the piece of plate to both the stud and in the bottom plate on both sides, as shown in Figure 6–14. Work the other side of the jamb just as any other jamb. Once the jamb is shot down and the door and king studs are screwed off, add the header and cripples to complete the jamb.

Window Jambs

You'll set the window jambs in much the same way as the door jambs. We'll begin by assuming the bottom and top plates are shot in place and laid out. Like the door jambs, the window jambs are fastened to the framing with clips. If the clips aren't welded into the jamb, you'll twist them into the window jamb just like a door jamb.

With the window jambs, it's important to maintain a constant elevation throughout the work area. In most cases, you'll line up the tops of the window jambs with the tops of the door jambs. Of course, shooting in benchmarks takes

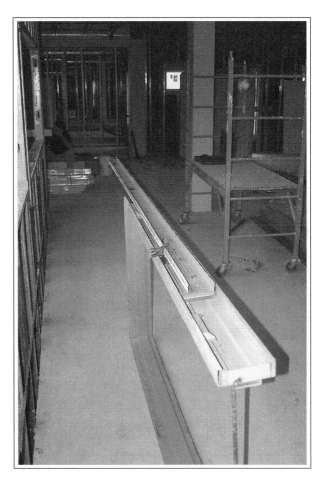

Figure 6–12. The door stud is being screwed off to the door clips before the door jamb is shot down. Hold up the bottom of the stud above the bottom of the jamb enough to get your hand and shotgun in, and to prevent the kick of the shotgun from jamming your hand into the bottom of the stud.

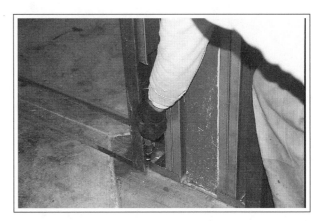

Figure 6–13. Reaching inside the jamb with the shotgun to shoot down the door cleat. The shotgun will kick; make sure you don't get your hand between the shotgun and stud.

Figure 6–14. Scabbing a piece of plate onto the door stud to fill the void will prevent a weak spot in the framing.

extra time. While some foremen will demand this attention to detail, others will get equally upset if you spend the extra time. As always, you'll just have to ask what a particular foreman or job requires. But this is the right way to do it, if the job situation allows it.

Step 1: Set the Window Studs The window studs, like the door studs, are the metal stud equivalent of the trimmer studs in wood framing. Get the window studs stood up on layout, and screwed off in the bottom plate. Then, measuring up from the floor or from a benchmark, establish your jamb elevation on at least one of the window studs. When framing in a window jamb that's close to a door jamb, use your level to set the top of the window jamb at the same elevation as the top of the door jamb.

Now you're ready to set and clamp the jamb in place to the window studs. Spread the window studs enough to let the window jamb slide between the studs, then fold the studs back tight to the clips and adjust the jamb to the proper elevation. This is easier to do with you on one side of the jamb, and your partner on the other. Clamp two clips on each side of the jamb to the window studs (Figure 6–15).

Center the jamb on the window studs so the gap between the stud face and the inside lip of the jamb is equal on both sides of the stud, just as you did with the door jambs. Again, the gap should be approximately the same width as the drywall that will be hung on the wall later. Once you've set the gap, set one side of the jamb to the desired elevation then level the jamb with your 4-foot level. Finally, screw off each clip with two 34s.

Step 2: Header It In As one partner cuts the top and bottom headers to length, the other will move the scaffold into position. Then both partners frame in the top and bottom headers simultaneously. Slip the headers in place between the window studs, and slide the bottom header up tight to the clips. Unless you're using clips in the top of the jamb, slide the top header down onto pieces of scrap drywall laid across each end of the jamb. Get the headers screwed off to the window studs, using one screw on each side of the stud on each shoe. On the bottom header, double-check the gap between the header and the lip of the jamb, and screw off each clip with two 34s.

Figure 6–15. The window jamb is set to the correct elevation and clamped to the window studs. The gap between the lip of the window jamb and the stud has been set, and the stud screwed off to the clips.

As soon as you slide the top header in place, get a measurement on the length of the top cripples. The partner working the bottoms should stop working the bottom header to cut the cripples for the top header to length, then finish up the bottom header.

To establish the layout for the cripples, measure from the hard side of a window stud to the center of the closest layout stud, at the bottom plate. Then transfer this number to the headers both above and below the jamb. Burn the number past the hard side of the same window stud and mark layout across the headers.

Get all your cripples stuffed up and screwed off on layout both above and below the window jamb, as shown in Figure 6–16. To make sure large jambs are sitting square, measure the jamb diagonally corner to corner, first from the top right corner down, then from the top left corner down. If the jamb is square, the number will be the same from both directions. Set your king studs and screw them off, and this jamb's finished.

Large and Lead-Lined Jambs

Large window jambs and lead-lined window jambs (which are really heavy) are framed in much the same way. The difference is that after you stand up the window studs on layout and screw them off in the bottom plate, go ahead and frame in the bottom header. Give yourself something to rest the jamb on as it's set. Set the header at the elevation for the bottom of the jamb, and make sure it's flat and level.

To install the jamb, leave the tops of the window studs unscrewed until the top header and cripples are framed in. Spread the window studs

Figure 6–16. With all your material in the area, it should take about 5 or 10 minutes to complete this window jamb in a 3⅝-inch wall.

and lift the jamb high enough to slip the jamb in between them. Some jambs will be too big and awkward to lift much higher than the header. When that's the case, unscrew the shoe on one end of the bottom header you've just installed so you can fold the window stud back out of the way. Once the jamb is set in place, fold the window stud back into position and rescrew the header. Then set the gaps and clamp the jamb in place. With the jamb secured, screw off the clips all the way around. Frame in the top header and king studs, and you're ready to move on.

Take a minute to step back and take pride in those jambs — strong, plumb and level. Then get ready for the next step. In the next chapter, we'll cover the step-by-step installation for furred walls.

CHAPTER 7

Furred Walls

In this chapter we'll take a look at three common materials and methods that are used to furr out walls on a commercial metal stud framing job. Both interior and exterior walls built from masonry and precast materials are often furred out instead of being framed with studs. This lets the framers build a quality metal stud wall while leaving as much usable room space as possible.

Because the framing methods used with these materials are nearly the same as ones we've already covered, I'll only go into detail on the drywall channel (DWC) method. And since Z-channel frames up about like DWC, and $1^{5}/_{8}$-inch (or *inch 'n five*) studs go in just about like any other metal stud wall, I'll focus mainly on the differences in the methods.

Each of these furring methods is a quick and simple way of framing up a wall. You don't need a top and bottom plate because the two furring strip materials are just laid out and shot to the masonry wall.

Furring with DWC

DWC (drywall channel), often called *hat channel* or *hat track,* is the most common of the furring materials. It's also used to frame in ceilings. The hat channel is available in two thicknesses: 7/8 inch, which is what you'll use in most situations, and 1 5/8 inch. Both are 2 1/2 inches wide and come in lengths of 10 and 12 feet, in both 20 gauge and 25 gauge material.

Instead of framing up the wall in a top and bottom plate, as we've done with all of the other walls we've covered, furring a wall with DWC requires only that you pull the layout and shoot the DWC to the masonry or precast concrete walls. In this example, we'll furr out a masonry block wall using 10-foot sticks of DWC.

Step 1: Establish and Pull Layout As with most walls, the layout of the DWC generally depends on what the drywall will butt against. This isn't always the case, though. You'll also furr out walls that'll share a common wall line with, say, a 3 5/8-inch stud wall. In this situation, the layout for the hat channel will follow the stud layout, or vice versa.

Establish the layout for the first stick of DWC on center, but because the DWC will cover it up the center marks, you'll establish the edge of the DWC and pull layout from there. Remembering that the DWC is 2 1/2 inches wide, measure 1 1/4 inches to one side of your center mark and make another crow's foot to establish the edge of the DWC. Mark an *X* on the side of the layout mark that the DWC will sit on. If you're wondering which edge of the DWC to start with, it doesn't really matter. I pull my DWC layout just like my studs. If you're working right to left on a stud wall, you'd mark the right hand side (soft side) of the stud, and set the stud to the left of your mark. Lay out your DWC the same way.

With the layout established, plumb your crow's foot up the wall using either a 4- or 6-foot level. You'll have to plumb up your line in sections, marking plumb up the side of the level, then resetting the level to the plumb line and continuing the line on up the wall. After you've plumbed your line up as far as you can reach, you're ready to pull layout.

Have your partner hold the end of your tape on the plumbed layout line about 16 inches or so up off the floor for the bottom layout marks. This has two advantages. First, you don't have to stoop over to pull layout (and your back will thank you). Second, it makes the layout marks easy to read as you shoot the DWC to the wall. Pull the top layout marks across the wall about as far up as you can reach. Pull both the top and bottom layouts across the wall as far as you can without your tape sagging (about 12 feet). The masonry or precast walls will really eat up a pencil. You'll get the best results by using a black felt-tip marker.

Step 2: Shooting It Off As you prepare to shoot the DWC to the wall, get your shotgun ready before you start. Be sure it's reasonably clean. If it's not, take a few minutes to clean it. Most powder-actuated nail sets are held together with just a couple of clips, as you can see in Figure 7–1. As you reassemble the gun, switch out the standard head for a hat channel head, which is set aside from the standard head in Figure 7–1. The thin end of the hat channel head

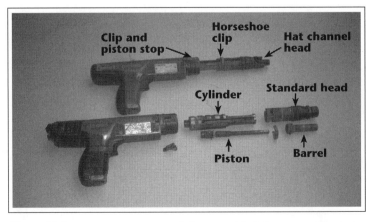

Figure 7–1. Here you can see a fully-assembled shotgun with a hat channel head in place of the regular head (top). Under it is a powder-actuated nail set disassembled, arranged in the order that it goes together. The entire gun is held together with two clips, the horseshoe (at the end of the barrel) and the clip that holds the piston stop in place.

is specifically designed to get into the thin edge of the DWC. There are other tight spots where the hat channel head will come in handy as well. However, the thin head will wear out pretty quick, so don't use it when you don't need it.

A gas-actuated nail set (Figure 7–2) won't require all this attention. The gas burns clean and leaves no residue like gunpowder, and the head is multipurpose so you don't have to change it. This saves a lot of time spent cleaning a powder-actuated nail set. The gas cylinder lasts for hundreds of pins and the magazine holds 30 pins at a time. All this adds up to an ideal piece of equipment for this and many other jobs.

In either case, you'll need to come up with two shotguns or nail sets if you can. If you can only get one, your partner will get your DWC leaned up against the wall out in front of you, then get your scaffold ready to run the tops. Once finished, your partner will start the prep work on your next work area until it's time to roll your scaffold again.

Begin the wall with a stick of DWC set horizontal along the floor, shooting it to the block wall, as shown in Figure 7–3. Set the edge of your first stick of DWC to the layout marks covering the X. Shoot the first two pins off right at the layout marks, with the edges of the DWC set exactly to them. Once it's tacked in place, shoot off each stick as far up as you can comfortably reach, working the tops off of your scaffold or a walk-up (a folding aluminum sawhorse).

Shoot the pins close to the mortar joints to prevent the pins from blowing through the block or splitting bricks when you're shooting to a brick wall. Space the pins approximately 24 inches apart, staggering them from side to side, as shown in Figure 7–4. To be sure the pins are holding the DWC tight to the wall,

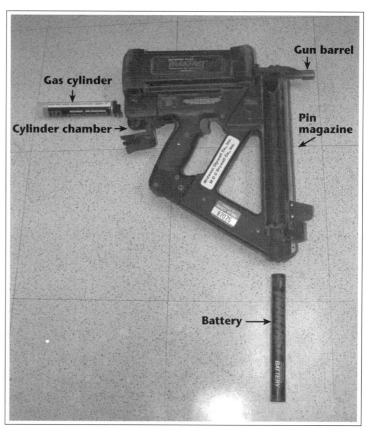

Figure 7–2. This is a gas-actuated nail set with the battery and gas cylinders pulled out in line with the sleeves they fit into. This tool is great for shooting to block, but it has its limits. Hardened concrete and steel are out of this tool's league.

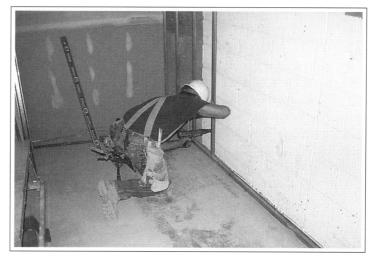

Figure 7–3. With the bottom stick of DWC shot in place horizontally along the floor, the first of the wall furring strips is set on layout and shot to the block wall.

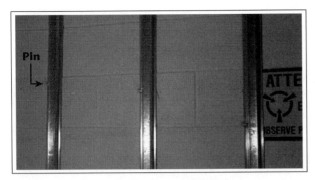

Figure 7–4. This shows the right and wrong way to shoot the DWC to the block. The pin on the left is correctly shot to the solid part of the block. The pin in the middle is shot into the weak portion of the block. The pin on the right is shot right into the mortar joint. This will hold the DWC solid but will pull dips in the DWC.

use your axe handle and tap any questionable area of the DWC. If it's loose, you'll hear it rattle.

Once you get the bottoms of the DWC shot off down the entire wall, hop on your scaffold and have your partner roll you down the wall as you shoot off the tops. This is where two shotguns really come in handy. With two guns, one carpenter sets and shoots off the bottoms while the other follows right behind shooting off the tops. You can see the wall being completed in Figure 7–5.

Step 3: The Corners As you shoot the DWC in place, you'll need to be aware that the outside corners of the masonry wall can blow out under the force of the shotgun. To prevent this, set the corner sticks of DWC 4 to 6 inches back from the outside corner, as shown in Figure 7–6.

Because there are obviously no sliders on a DWC wall, the outside corner will be completed as the wall is rocked. As you rock the first side of the corner, run the rock wild past the corner at least 6 inches and screw off this sheet completely. Now straightedge the wall line of the intersecting wall over to the brown side of the rock in two places and snap a chalk line (Figure 7–7). Complete the corner by cutting back the drywall and screwing either a piece of wall mold (grid ceiling material) or utility angle to the new corner from the front side of the rock with drywall screws spaced 8 inches apart. See Figure 7–8.

The inside corners are considerably less trouble. First, shoot a stick of DWC tight to the corner point on the first side of the corner to be rocked. Set the second stick back just far enough to get the hat channel head in to shoot off the edge of the DWC. It's very important to properly frame both the inside and the outside corners so they're solid. However, keep in mind that this is

Figure 7–5. The DWC wall being completed by a carpenter using a walk-up to reach the tops of the DWC.

Figure 7–6. The DWC is held back approximately 4 inches from the outside corner of a concrete column to prevent blowing out the corner. Finish the actual outside corner off with a stick of utility angle.

Figure 7–7. Using a 4-foot level to straightedge the actual corner point over to the back side (brown side) of the drywall that forms the opposite side of the corner. Transfer the wall line over at the top and bottom of the wall, then snap a line and cut off the drywall.

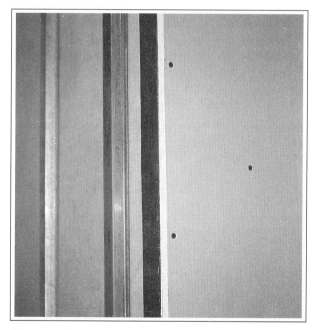

Figure 7–8. With the drywall cut back to the exact corner point, simply screw the utility angle off to the drywall, following the cut edge.

only important below ceiling height. Make your corner solid to about 6 inches above ceiling height. From there on up (as far as the wall goes), you can float the corners.

DWC on Concrete Columns

Framing around concrete columns is another place where you can use DWC. Simply furr out the four sides of the column instead of framing around it with studs. Then just set the DWC to the column 4 inches in from the outside edges. No layout is required. Just plumb each stick of DWC with your level (or by eye once you develop the knack) and shoot it in place. Space the pins 24 inches apart, staggering them from side to side.

When you begin furring out the columns, you'll quickly notice how hard the concrete is. After all, it's poured with structural integrity in mind. The concrete is often so hard that you'll have to switch to steel pins, because the concrete pins simply bend in half or shear off.

The pins shearing off can cause problems all their own. A sheared-off pin is going to do one of two things: jam up the gun, or ricochet off the concrete and fly who knows where at high velocity. When a pin jams the gun, find a couple of scrap pieces of 2 x 4 and stack them up. Then milk the shotgun down by pressing the barrel into the firing position and releasing it a couple of times. Now shoot the sheared-off pin into the 2 x 4. These are also common problems when shooting to steel. The only thing you can do is keep the gun as square as possible with the surface you're shooting to, and wear your safety glasses.

Another condition that's common to the concrete columns is how incredibly out-of-square and uneven they can be. Although each job is different, generally, if the prints call for the columns to furr out, that's what you'll do. But when the column is really bad, you'll be ahead of the game if you switch to $1^{5}/_{8}$-inch studs. Of course, until you've established yourself with an outfit, this is a call your foreman needs to make.

Furring with $1^{5}/_{8}$-Inch Studs

There are two situations when you'll use inch 'n five studs to furr out walls. The first is when $1^{5}/_{8}$-inch studs are called out in the prints. The second is when you have to use them instead of DWC to square up a room, or when the pre-existing wall is too difficult and slow to shoot to. The $1^{5}/_{8}$-inch studs are 20 and 25 gauge and come in several lengths, which will get you over ceiling height in most situations.

The 1⅝-inch walls frame up pretty much the same as any other metal stud wall we've covered, especially the freestanding walls. The biggest difference is in how you brace the 1⅝-inch walls, so that's where we'll focus. For this example, we'll frame up an 8-foot wall with the bottom plate shot down and laid out. Because this wall will frame up just about like a freestanding wall, I'll skip the step-by-step discussion and just highlight the differences.

Top Plate

With the bottom plate shot down and laid out, stand up a stud on layout in each end of the first stick of bottom plate and slide your first stick of top plate onto them. To support the top plate, use either a gusset or a simple stud brace. Here we'll use the stud brace. Plumb the studs by eye, then lift the top plate up off the studs about ⅜ inch. Now just set the brace on top of the 1⅝-inch top plate and shoot it near the mortar joints in the masonry. See Figure 7-9. Space the braces approximately 4 feet apart down the entire length of the wall. If you have any problem keeping the joints in the top plate straight, set the wall to a dryline later and add a brace at the joint (or joints) that are giving you problems.

Bracing the Studs

You'll stuff up and plumb like the freestanding wall, then set it to a dryline at the top of the wall to straighten it. What we want to look at here is the fact that only one side of the wall will be rocked, taking away much of the rigidity of this (or any) light gauge wall. To increase the wall's rigidity, brace the studs to the masonry wall. You'd need to do this regardless of the width of the studs, but the 1⅝-inch studs need more braces.

To brace the wall, the simple stud brace is the quickest way to get firm, rigid results. You'll need to consider the gap between the masonry wall and the studs of the wall you're framing. Unlike the braces at the top plate, which are above ceiling height and can stick out past the framing, the stud braces obviously can't. The ideal situation is to have the brace hitting at about the middle of the stud, as shown in Figure

Figure 7-9. The top plate of a 1⅝-inch wall is interrupted for some large conduits, with a stud brace on each side. The braces were set (more importantly) closer to the plate than the mortar joints. The shotgun was milked down to reduce the force of the shotgun so it didn't blow through the block.

Figure 7-10. Simple stud braces shot to a precast concrete wall at about the middle of the 1⅝-inch studs, then screwed to them with two framing screws.

7-10. This may mean you have to cut your braces out of a different width of stud, or you may need to use gusset-style braces for the wall. The braces will do the best job when they're shot to the masonry about 4 feet off the floor. As always, each job is different, but bracing every other stud is usually plenty.

Here's a shortcut that's often used to brace the studs: shove a scrap piece of drywall about 12 inches long tight to the masonry wall and screw it to the hard side of the studs (Figure 7-11). It doesn't matter how wide the drywall is to start. Just let it run wild out past the face of the studs

Figure 7–11. A 1⁵⁄₈-inch wall stiffened up with a piece of scrap drywall. This little trick will save you lots of time.

and then cut it back. Be sure you rasp the cut edge of the rock smooth. This is a great time-saver, but I don't recommend using this type of brace alone. Without some kind of stud bracing, the drywall braces will let the wall bounce. That will cause the joints in the drywall to crack later, after the wall is rocked and taped.

Furring with Z-Channel

The Z-furring channel wall is framed about the same as a DWC wall, requiring no top or bottom plate. The difference is that the Z-channel is used specifically on exterior walls, so there's room for the insulation that's used with the Z-channel furring system. The Styrofoam insulation used with the Z-furring system is available in widths of 16 or 24 inches, to work with either layout spacing. The Z-channel is a 25 gauge material that comes in four depths, ranging from 1 to 3 inches. It gets its name from its shape (Figure 7–12).

As you can see in Figure 7–12, the two flanges of the Z-channel are different sizes: one is ⁷⁄₈ inch and the other, 1¹⁄₄ inches. You'll use the ⁷⁄₈-inch flange to shoot the Z-channel to the wall, while the drywall is screwed to the 1¹⁄₄-inch flange. That 1¹⁄₄-inch flange also holds the Styrofoam insulation in place. Like the studs, all the Z-channel will face the same direction. That leaves only one side of the insulation held in place, but it'll usually do the job until the wall is rocked. When conditions require it, you can lace wire through the punchouts in the Z-channel and over the insulation to hold it in place.

Step 1: Establish Layout   Like all walls, the layout for the Z-furring channel depends on whether the rock will butt to the rock of an inside or outside corner, or into an existing condition, like the red iron or masonry. When the rock butts to an existing condition, simply establish the center for the first stick of Z-channel off of that condition. From the center mark, establish the edge of the ⁷⁄₈-inch flange, measuring an additional 1¹⁄₂ inches (⁷⁄₈ inch for the flange plus ⁵⁄₈ inch for half the width of the 1¹⁄₄-inch flange). Then plumb this point up the exterior wall with your level and pull layout from this point.

When you'll butt to a corner, measure off the depth of the Z-channel you're using in two spots close to the corner of the block wall. Do it on each side of the corner. Draw out the corner using a straightedge, then from that corner point establish first the center, then the flange of the Z-channel. Plumb this point up the block wall

Figure 7–12. A stick of Z-furring channel as it would sit against a concrete or masonry wall as it's shot in place. The slotted grooves running through the center of the channel help to relieve the pressure when the Styrofoam insulation is installed as well as letting the wall breathe.

and pull layout from here for both the top and bottom of the Z-channel. At the outside corner of the block wall, establish the layout marks 6 to 8 inches back from the corner and transfer them up by measuring off the other layout marks.

The rest of the wall is clear sailing. Work your way down the wall setting the 7/8-inch flange to the layout marks and shooting it off, as shown in Figure 7–13. The insulation will be installed after the Z-channel is shot in place. You'll see other carpenters actually use the insulation to set the layout for the Z-channel, installing the insulation as they frame the wall. This method is acceptable on some jobs, but I don't recommend it. The insulation isn't consistently the same width, which creates an inconsistent layout and potential problems in later steps of the work. In Figure 7–14, you see a properly laid out and shot off Z-channel wall.

Step 4: The Insulation With the framing completed well out in front, your partner will begin installing the Styrofoam insulation. Few cuts are made to the insulation. It simply slides under the flange of one stick of Z-channel and bumps up against the back of the next stick. The insulation will simply stay wedged in place. Figure 7–15 shows a completed wall.

Resilient Furring Channel

The resilient furring channel seen in Figure 7–16 is attached to the metal stud (and wood) framing horizontally to control sound. The *RC channel*, as it's commonly called, is fastened to the metal stud framing on only one side of the channel to allow the walls to float. The RC channel acts as a bridge between the wall studs and the drywall, preventing sound transfer from room to room.

Step 1: Establish Layout The standard spacing for the RC channel is 24 inches on center. Like the DWC and Z-furring channel, you'll pull layout for the flange of the channel. Begin by measuring up off the floor 24 inches at each end stud. The RC channel is 2 1/2 inches wide, but 1 inch of it is the flange and a break point in the material. Add the 1 inch to 3/4 inch (half the width of the face of the channel) for a total

Figure 7–13. The Z-furring channel shot in place on layout with the sticks of channel facing the same direction.

Figure 7–14. A completed Z-furring wall.

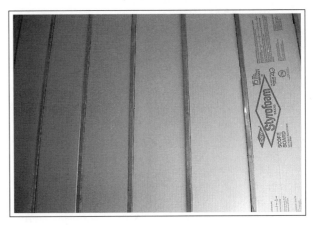

Figure 7–15. The Z-furring wall with the Styrofoam insulation installed.

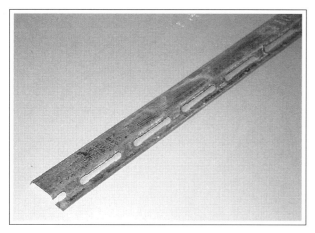

Figure 7–16. A stick of resilient channel. You screw the skinny edge of the channel to the wall studs.

Figure 7–17. The first row of resilient channel along the floor with the flange turned up. The flange is turned down on the next and remaining rows, with the RC screwed to each wall stud.

of 1³/₄ inches. Measure down from the 24-inch marks 1³/₄ inches to establish the flange of the RC channel. From these points measure up the wall, marking the layout at 24 inches on center. Next, snap a chalk line between the end studs to mark the layout to all the studs.

Step 2: Screw the RC Channel to the Studs Set the bottom stick of RC channel to the studs with the flange up. You can just eyeball it about ¹/₂ inch up off the floor. Screw the channel off to each wall stud with one framing screw. Set the rest of the channel to the layout lines with the flange down, again screwing the RC channel to each stud with a single framing screw. In Figure 7–17 you can see a 3⁵/₈-inch metal stud wall furred out with RC channel.

Well, that's about it for framing the interior with light gauge metal studs. In the next chapter we'll start learning about structural studs.

CHAPTER 8

Structural Stud Walls

In this chapter, we'll take a look at the differences in both materials and techniques in framing with structural studs. Overall, the methods you'll use for framing with structural metal studs aren't a lot different from framing with the light gauge interior studs we've already covered. That's why I'll go into a detailed step-by-step discussion only for bearing walls. Then I'll cover the methods for various other structural wall conditions that are different from what we've already learned. A couple of these sections will center around fastening the tops of the studs in place, which isn't always into a top plate, as you'll learn.

In previous chapters we looked at framing with light gauge studs, which range from 25 gauge to 20 gauge. The structural studs range from 18 gauge (the lightest), through 16 gauge, 14 gauge and 12 gauge (the heaviest). There's also a 20-gauge structural stud. It's pretty easy to tell the different gauge studs apart, since they're marked in two different ways. First, each stud is color-coded with spray paint:

- ◆ 20 gauge is white
- ◆ 18 gauge is yellow

Commercial Metal Stud Framing

- 16 gauge is green
- 14 gauge is blue
- 12 gauge is red

The structural studs are designed to do specific jobs in supporting the building. They're commonly used to frame up solid structural weight-bearing walls to support the roof or deck above, on both interior and exterior walls. The structural studs are also used to frame up exterior walls with deflection plate or deflection clips (or *slide clips*) to carry the tops of the studs. The deflection plate and clips allow the roof to drop a little with the added weight of snow and ice, while maintaining the walls' structural integrity to stand up to wind shear.

When working a structural wall, you'll immediately notice some differences in the basic framing methods. Cutting the studs, for example, will require a chop saw or Quickie Saw (a chain saw with a chop saw blade). A torch comes in handy for making oddball cuts. In a pinch, you can use the scoring method covered in Chapter 1 to cut the studs. This isn't very productive, but it'll keep you working.

The screws you use to fasten the studs into the plate will change as well. The heavy gauge material requires self-drilling framing screws (commonly called *S-12s, lath screws* or *34s*). Now let's get to it.

Bearing Walls

For this section, we'll walk through the step-by-step methods you'll use to frame up a weight-bearing structural stud wall. It's 11 feet tall, using 12-foot 16 gauge studs. As with any job, you'll first get the area cleared and stocked, set up your equipment and put the scaffold together.

Step 1: Bottom Plate The structural stud walls are plated basically the same way as any of the other walls we've covered. The biggest difference is that you don't have to splice the bottom plate together. Instead, just butt the ends of the plate together and shoot it in place. Splicing the heavy gauge bottom plate would be counterproductive.

The exterior walls are often framed in sections that run between the red iron piers (columns) of the superstructure. This buries the piers inside the wall. In this case, there's no need to notch the plate. Just run your plate up to about 1/4 inch short of the piers and shoot it down following the wall line. Pay close attention to the red iron to be sure it's jibing with the wall line, which is a common problem. Ironworkers often have trouble with straight lines and the concept of plumb.

Don't let the fact that the pier's set inside the wall line 1/2 inch or so fool you. The piers will often be wrapped with one to four layers of drywall to fireproof them. All this must fit inside the wall or it'll cause large bumps in the drywall later. Any problems you deal with now will prevent three times the problems later.

As with all metal stud walls, the bottom plate will establish the corners of the structural exterior walls. Simply shoot the two pieces of plate that form the corner down separately. Don't leave a gap for the rock to slide past. When framing with structural studs, frame a hard corner using a stud at each corner point, as shown in Figure 8–1.

There's one more situation you're likely to run into. Sometimes the concrete hasn't been poured around the red iron piers. When you run into that, just let your plate run wild out over the void, as shown in Figure 8–2. Once the concrete is filled in and allowed to cure, you can go back and add a couple of pins later, if they're needed.

Step 2: Plumbing Up You'll have to plumb up the structural stud walls just like any other wall, but there's an additional complication—the wind. To overcome it, a laser is the ideal solution. But if you can't get your hands on one, a

Figure 8–1. You can see a separate structural stud forming each individual corner point both inside and outside, just like a corner formed in wood framing.

Chapter 8: Structural Stud Walls   117

Figure 8–2. A typical concrete "diamond" is left unpoured around a red iron pier, allowing the pier to be set and adjusted. The 14 gauge bottom plate runs far enough out over the void to set the two window studs.

heavier 16- or 32-ounce plumb bob will do the job. If the wind is so strong it's hard to steady the plumb bob, stand a stick of plate right next to the plumb bob and string line to block the wind. You'll need to get someone (preferably from your own crew) to hold the stick of plate for a minute as you plumb up the wall line, as shown in Figure 8–3. Using either method, get the wall line plumbed up and snapped out. The red chalk will usually show up fine against the red iron, but if you have any problem seeing the line, use a blue chalk line.

There's another problem that's common to plumbing up to the red iron. Once you've plumbed up your wall line, the outside of your wall won't clear the steel (Figure 8–4). In most cases, the exterior wall will need to at least run flush with the steel to allow the exterior gyp board to run past and cover the steel. While this will present a problem, there's usually a simple solution. Your foreman has to make the call here, not you, but it's common to frame up the wall out of plumb, simply following the outside edge of the red iron. For this example, however, we'll assume this isn't an issue, and just let the top plate follow the chalk line.

Step 3: Shooting Up the Top Plate   As you shoot the top plate up into place, use the 90-degree plumb mark to establish the stud layout and pull the stud layout as you go. This lets you check the stud heights, which often fluctuate a lot. Then have your partner start cutting the studs to length.

The top plate of this wall is being shot to the red iron. Remember that you'll have to use steel pins whenever you're shooting to steel. The stubby steel pins, which are only $3/4$ inch long, often get mixed in with the concrete pins, the

Figure 8–3. Using a 6-inch stud to block the wind while plumbing up the wall line from the bottom plate to the deck.

Figure 8–4. This is a stick of 6-inch 14 gauge plate set to the top plate line, plumbed up to an I-beam. With the plate clamped in place, you can see that the exterior wall line won't clear the red iron. Here we'll slide the plate flush with the I-beam, leaving the wall slightly out of plumb. That allows the wall to clear the steel.

118 Commercial Metal Stud Framing

shortest of which is 1 inch. You need to be able to quickly distinguish the two types of pins, shown in Figure 8–5. You can shoot steel pins into concrete, but don't shoot concrete pins into steel. They'll shear off. Use the same 24-inch pin spacing as you used on the bottom plate.

Step 4: Stuff It Up Since we're framing an 11-foot wall using 12-foot studs, cut the studs to length with only 1/8 to 1/4 inch leeway off of the exact number. The weight of the deck above must ride on the studs, *not* the framing screws holding the studs and plate together. Any time you're cutting structural studs for a bearing wall, cut the studs as tight as you can without having to fight them into the plate. Your partner will start cutting the studs as you're shooting up the top plate. You can also give a number measuring right up to the I-beam before the first stick of top plate is even shot up. Cut the studs to length on the chop saw. As always, you'll measure the stud length from the same end of the studs every time so the stud holes line up. Even when cutting 8-inch studs, you'll be able to cut and stand up several at a time, with the cut end up on each stud.

Next, set the studs on layout and get them screwed off at the front and back of both plates. It's also common on weight-bearing walls for the studs to be welded to the top plate as well as shot up. It's a plus if you have some welding skills, and it certainly increases your earning potential.

Cutting the structural studs will eat up a chop saw blade pretty quick. The heat generated also causes the blade to glaze over, which prevents the blade from cutting properly. Whenever you're using the chop saw for cutting structural studs, grab an extra blade out of the gang box, and keep it close at hand. There's no right or wrong way to put a blade on, just be sure the bolt (which has reverse threads) is good and snug. For obvious reasons, *always unplug the chop saw when changing blades.*

Step 5: Lace the CRC In most cases, you'll lace the studs of the structural wall with CRC, which is also called *black iron*. In our sample wall, we're framing it in sections between the red iron columns. That's a good example of a wall that should be laced as it's stuffed up. Lace the CRC through the studs every 4 feet, beginning at the row of stud holes closest to 4 feet off the slab. Once you have several studs stood up and screwed off, slide enough sticks of CRC through each of the appropriate rows of stud holes to lace the entire wall, as shown in Figure 8–6. After you get all the studs in this section of the wall stood up, pull the CRC down through the rest of the studs (Figure 8–7), overlapping the joints of the CRC by at least 16 inches — or one stud hole each way.

Figure 8–5. A typical 1-inch concrete pin on the left and a steel pin (3/4 inch) on the right.

Figure 8–6. Here are several 6-inch structural studs stood up on layout, with the CRC for the rest of the wall loaded up on them.

Just as with the studs and the top plate, the CRC is often welded to the studs. This requires little more than a tack, but it needs to be a solid tack weld.

Step 6: Rough Openings Let's take a look at the rough openings (R.O.) for the doors and windows. We'll begin with an exterior window R.O., which won't normally use a hollow metal jamb. Most exterior windows have wood jambs. Your first move is to plumb up the layout mark for one side of the R.O., which is laid out along the bottom plate line. Then you can measure the other side of the R.O. from the plumbed side. Next, stand up the window and king studs and screw them off squarely in the top and bottom plate. Cut the bottom header with ears, and install it level at the bottom R.O. elevation. Finally, install the bottom cripples on layout and screw them off. See Figure 8–8. The cripples cut for a structural stud wall will often require a little more thought and work than in other walls. We'll look at that process in more detail a little later.

The top header of a window rough opening or over a door jamb will require a little more work than the bottom. Any time the top header of a structural wall will bear any weight, use a box beam header. Your partner should begin building the needed box beams well before you get to the rough openings. Use the layout marks of the R.O. along the bottom plate to determine the box beam's length, which is the only number you need. Since we've already covered building the box beams in Chapter 2, let's move straight into the installation process.

First, establish the bottom elevation of the box beam by measuring up from the bottom header or the slab, and then leveling across if the bottom header's not framed in yet. You've always got to focus on the high work, and your partner on the ground will concentrate on keeping you supplied with material. Then gang up on the bottom work later. Tilt the box beam diagonally between the studs and slide the tabs of the box beam over the window studs as you tilt it back into the level position. Lower the bottom of the box beam to the desired elevation and clamp it off, setting first one side then the other (Figure 8–9). Check the R.O. to maintain the width, and make any needed adjustments by tapping the box beam up or down with your

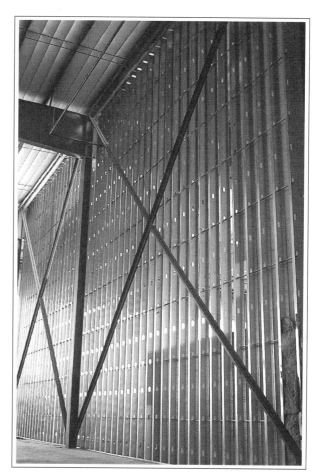

Figure 8–7. A 6-inch structural stud wall, with the CRC laced through the stud holes every 4 feet from the floor to the deck.

Figure 8–8. Here's a header with ears cut from structural plate and screwed off to the window studs. The cripples are screwed off on layout.

Figure 8–9. A typical box beam header set at its correct elevation and clamped in place. Look closely and you'll see that this box beam was insulated as it was built.

axe. Once you've got it where you want it, screw off each tab of the box beam with three lathe screws.

The Cripples

When CRC runs through the studs, you'll often have to double cut the cripples so the stud holes will keep lining up above and below the rough opening. This is particularly important on tall walls that have several rows of CRC running through the cripples above the R.O.

To line up the holes, simply measure up from the top plate of the box beam to the bottom of the first hole in one of the window studs. As you cut the cripples to length, first measure from a stud hole at one end of the stud and mark your number. Then measure from this point and mark the cripple length. You'll make two cuts on the cripples: one cut will match up the stud holds, while the second will cut the cripple to length.

Once cut to length, install the cripples on layout and screw them off. Then slide the CRC you loaded up in the wall studs through the cripples. Obviously the CRC won't run through the rough opening. The rows of CRC that run at the same elevation as the R.O. will run through the king stud and stop at the holes in the window studs.

The rows of CRC won't always line up to run under or over the rough openings, but the cripples may still need the added strength of the CRC. In this situation, cut a piece of CRC long enough to span the rough opening plus at least one layout stud on each side. Then, before you add the cripples, load the CRC into the stud holes of the wall studs closest to the headers, both above and below the rough opening, as needed. Now frame in the cripples and slide the CRC through them.

Doors

For typical entry doors, you'll set and frame hollow metal jambs following the methods outlined in Chapter 6. The only difference is that on a structural wall, the header will be a box beam and CRC may be required to run through the wall and cripple studs.

In many situations, especially the main entrance of a building, the doors will be part of a large glass and mullion wall that will be installed in one piece in your framed rough opening. Here you'll often find yourself working with the glaziers, who'll need only to be able to set their mullion plumb, level and square. Few pieces of work on any job are set truer than the mullions, and you'll often frame walls whose placement is determined by the vertical members of the mullion frames.

The most important factor you need to watch for when framing the rough opening for a mullion is to make sure there's plenty of room for adjusting the mullion. When you establish the header elevation on large rough openings where the mullion will sit directly on the slab, check the concrete for level across the opening. Measure the header elevation up from the high side of the concrete and level it across the rough opening. This will ensure there's room to level the mullion, and the glaziers will shim up any gaps. In short, the rough opening is much better framed too big, than too tight.

Structural Stud Wall with Slide Clips

Another common use of structural studs is framing in exterior walls that must stand up to the elements while allowing the roof to lower and raise (deflect) with the weight of snow and ice. In Chapter 1 we covered how to frame a wall using deflection plate, and you follow the same procedure when you're using deflection plate on a structural wall. The only differences are that the deflection plate is heavier gauge and it's usually shot to the red iron of the building superstructure.

Here we'll cover the material and methods used to frame a structural deflection wall using slide clips (shown in Figure 8–10) instead of deflection plate. The slide clips are set against the inside leg of the studs and then welded to the red iron. This technique gives the wall its structural integrity and allows deflection, while letting the studs run up past the roof line to form a parapet wall.

Step 1: Establish the Elevation The first step is to establish a constant elevation, because the tops of the studs will be plated later to form the parapet wall, which must be level around the building. On most commercial projects the General Contractor will have benchmarks shot in around the job as soon as the red iron piers are set throughout the concrete pad. If this isn't the case, your first move is to set up a laser in the middle of the pad and shoot in the benchmarks at all the exterior piers.

With the benchmarks shot in, you can start to figure out how much work it'll take to maintain a fairly constant stud height. You'll start by leveling the benchmarks around the piers. Then measure from the bottom plate to the benchmark on the piers near the corner of the building where you're starting the wall. As long as the difference between the benchmarks and bottom plate isn't more than $3/4$ inch, you can cut all those studs the same length. When this works out, figure the stud length at the benchmark from the shortest measurement between the plate and benchmark. This will help insure that the studs aren't too tall. You don't want to waste time trimming them.

When the pad is too uneven to use one stud length, cut and set a stud on layout closest to a benchmark at each bay and pull a dryline between the tops of the studs. Cut the studs $1/2$ inch short of the top plate height, then set your dryline to the inside of the studs 3 inches short of the top plate height. Leave the dryline up and use it later to set the top plate. To determine the height of the studs in between, measure up to the dryline and add $2 1/2$ inches. That gives you $1/2$ inch play when setting the top plate.

Step 2: Establish and Pull Layout To save a little time here we'll assume that the bottom plate is already shot down and laid out, so your next move is to plumb up the layout. When working a wall like this, you'll work the tops from up on the roof, which will require you to wear a harness and lanyard. Because this work is outside in the elements, and you'll be plumbing up 15 to 20 feet or more, use the heaviest plumb bob you can find. Many carpenters have custom plumb bobs, weighing as much as a pound, made just for this type of work. But the pocket laser I talked about in Chapter 1 is the best tool for the job.

You'll be plumbing up to the top piece of red iron at the roof line, which is commonly a piece of $3 1/2$- by $3 1/2$-inch angle iron. Plumb the layout up and mark it at the top of the angle iron using either your black magic marker or a soapstone. These exterior walls are often several hundred feet long. To maintain the layout, plumb it up from the bottom plate every 50 feet or so, establishing the soft side of the studs, and pull the layout from there. When you establish the soft side of the studs, check with your ground man and make sure you're both on the same page. Once established, set the end of your tape to the layout mark and use a pony clamp (spring clamp) to clamp your tape to the angle iron, and pull layout.

Figure 8–10. Five 6-inch structural studs fastened in place with slide clips, which are welded to the red iron.

Step 3: Standing the Studs When standing studs around the exterior of a building like this, whether they're fastened with slide clips or welded directly to the angle iron (which we'll cover next), the studs will often be more than 25 feet long. This will present some real challenges — cutting the studs for example. With studs this long, it's a lot faster and easier to move the saw to the material instead of the material to the saw. You'll still be able to cut several studs at once.

Standing the studs also presents an obvious physical challenge, but there's a simple solution. Use a length of rope with a C-clamp tied to one end to help stand the studs. We'll assume you're working the tops here, from up on the roof. Have your partner cut a stud to length, then lay it down perpendicular to the wall, close to where it will stand on layout. Make sure the bottom of the stud is against the foundation to keep the stud from sliding as you stand it up. Fasten the C-clamp through a stud hole at the far end of the stud. Next, as your partner walks the stud up from the far end, you'll help by pulling up the top of the stud with the rope.

Once the stud is standing, both partners can lift the stud at the same time and set it in the bottom plate. Then, while your partner is setting the bottom of the stud on layout, you'll set the top on layout and slip a clip on the inside leg of the stud. Then clamp the clip to the red iron, as shown in Figure 8–11. Stand and clamp off four studs (or as many you have ready) at a time, then weld only the slide clips to the red iron. See Figure 8–12.

While your partner is cutting the next few studs, you'll weld the ones that you've got standing and clamped off. It's usually all your partner can do to keep the studs coming. It's rare that he (or she) can also get the bottoms of the studs screwed off. That's okay, but they *must* be set on layout.

It's important to learn to develop systems like this quickly with all your work. Also keep extra welding rod, framing screws and chop saw blades close at hand, as well as a full tank of gas in the welder. Every time one of you has to stop and go get material, it breaks the rhythm of the work. Get set up, get working and stay working. That's the key to productivity. If your foreman has a laborer who can help just carry

Figure 8–11. Three 6-inch structural studs set in place on layout with the slide clips clamped off to the red iron angle.

Figure 8–12. The slide clip is set against the inside lip of the stud, holding the stud tightly to the red iron, and welded in place.

the studs from the saw and stand them up, your production level will go way up.

Don't forget the cold-rolled channel. If there are obstructions, like red iron columns, that'll interfere with lacing up the studs, you'll need to at least load up the studs. This is another task where an extra pair of hands will really help out.

If a wall requires CRC but you can't get it laced as you frame the wall, it's harder to lace it up. It'll take both partners; one forces the CRC through the first couple of holes while the other balances the end. Once you get the end of the CRC through the first few holes, it'll get a little easier. And the entire method is easier on 24-inch than 16-inch layout. Do yourself a favor and wear a pair of gloves when forcing the CRC into the studs. See Figure 8–13.

We've covered most of the basic methods and common problems associated with framing this kind of exterior wall. As you go from project to project, you'll run into other oddball situations that are just part of the trade. When you do, stay focused on picturing how the finished wall will look. If you can do this, you'll quickly discover that there are no problems, only solutions waiting to be found.

Structural Walls Welded to the Red Iron

Welding the structural studs to the red iron generally follows the same procedures as the slide clip method we just covered. The only difference is that you'll weld the studs directly to the angle iron. This allows the structural studs to support the weight of the roof while letting the studs run up past the roof line to form the parapet wall. With both of these methods, be prepared to frame the wall out of plumb. The superstructure will rarely line up exactly with the slab, which is necessary for the wall to be plumb. Again, your foreman and the General Contractor's superintendent will have to agree on what will be done — which is usually to frame the wall out of plumb.

If the method of welding the studs to the red iron isn't called out in the prints, your foreman will make the decision in the field. Once the studs are stood and clamped off on layout, they'll be welded either vertically up the two sides of the stud (Figure 8–14), or horizontally across the top of the angle iron. Some studs will have to be welded using both methods.

On nearly all commercial projects, welding the structural studs will require a certified welder. If you have strong welding skills and can run a good bead, you shouldn't have any problem. If you can't weld, take my word for it: Learn how, and you'll greatly increase your value to your outfit — and your long-term employability.

The biggest problem you'll have welding the studs in place is burning into the red iron without burning through the studs. You'll get the best results using a light rod with the welder set low, around 100 amps. As you run your bead,

Figure 8–13. Forcing the CRC through the stud holes of a 6-inch structural stud wall. Once the CRC is through the first hole, bend the CRC around to the next hole and push or pull it on through the studs.

Figure 8–14. You can see a structural $3^{5/8}$-inch stud welded to the red iron with a vertical weld up the stud.

let it burn into the red iron while lightly cutting into the stud.

Parapet Walls

In the last two sections, we left the structural studs standing around the perimeter of the building. The tops of the studs were run up past the roof line in order to form the parapet wall. Now let's take a look at a couple of different ways to frame a parapet wall. It's usually a 4- or 5-foot wall framed above the roof line to conceal the roof-top mechanical units. The first method we'll cover will complete the studs that ran up past the roof line. Then we'll look at framing a parapet wall when the exterior studs stop under the red iron, as they did earlier in the bearing wall.

With either method, your first step is, as always, to set up to do the job. Move your chop saw, screw guns and all your materials up on to the roof so you're not going up and down getting things. Most projects have at least one large forklift that you should be able to use to get your material up on the roof.

Step 1: Elevation In the slide clip method, I told you how to set a dryline that you'll use to measure the stud height and then again later to set the top plate. For this discussion we'll start from scratch, assuming no dryline was used. To begin, your partner will work down on the ground, using a 2-foot level to transfer the benchmarks from the red iron piers to the closest wall studs. Now, as your partner holds the end of your tape to the concrete slab, measure up the stud and mark it 3 inches short of the top plate elevation. Pay close attention to the prints when you're figuring the wall height; nearly all parapet walls are capped with double 2 x 6 or 2 x 8 material. Your wall height is at the *bottom* of the 2-by material, so you'll have to allow for it. Then, from your 3-inch mark, measure down to the benchmark and remember (or write down) this number. You'll use it to transfer the rest of the 3-inch marks up from the benchmarks, maintaining a constant elevation.

On smaller jobs, transfer up all the 3-inch marks and get it out of the way. On larger jobs, transfer up enough 3-inch marks to keep you busy for a couple days (and you'll be plating several hundred feet a day). Snap a chalk line to the studs between the 3-inch marks, which will be about 50 feet apart. Set all the top plate 3 inches above the chalk line to keep it level and flat.

Keep in mind that you use this method to set the top plate exactly level and true. Depending on what the parapet wall is capped with, this level of quality and work may not be necessary. Here again, your foreman will make the call. If he doesn't tell you what he wants as he gets you lined out, ask. You may be able to just slide the top plate onto the studs and screw it off, without setting it to a chalk line.

Step 2: Plating the Studs You'll begin plating the studs at a corner — and the plate itself will usually form the corner. If you recall from the slide clip method, you'll usually have to stand the corner studs as you plate the corners. Set the first stick of plate on top of the studs and eyeball it just shy of the outside face of the studs in the intersecting wall.

Slipping the plate down onto the studs is often a difficult job. Set the plate on the studs and let the back side of the plate slip over the back edge of the studs, then pull the front out. Slide the plate down onto the studs, one at a time, as shown in Figure 8–15. Work the plate from end to end. Once you get the plate down on a few studs, clamp them to keep the plate from popping back off. As you work the top plate, there's always the possibility that you'll drop tools or material off the roof. Keep an eye on the ground below and let others know you're working above them.

Once you slide the first stick of plate down onto the studs, use a straightedge and set the outside corner. Set the end of the plate 1/4 inch short of the actual corner point,

Figure 8–15. The framer is slipping the lip of the top plate over the outside edge of the studs, then pulling out on the inside lip of the plate at each stud and shoving it down onto the studs.

using your 4-foot level as a straight-edge (Figure 8–16). To adjust the plate to the corner point, remove any clamps and tap the end of the plate with your axe, hitting the plate right on the corner. Notice, I said *tap,* not *beat,* the end of the plate up. After you get the plate set, tack it in place with a couple of lathe screws.

After you've tacked off the first stick of top plate, your partner may go down to the ground and help you stand the corner stud. Then he'll set the bottom of the stud and get it screwed off. If you don't have a cordless screw gun, you can tie the screw gun cord and the end of the extension cord together and lower the screw gun to your partner.

For now, just tack the top of the stud to the outside of the top plate, then set it later when you're setting the plate to the 3-inch line. With the first stick of plate tacked in place, use the splice or scab methods we looked at in Chapter 1 to form the joints in the plate. At this point, you and your partner will split up, with one continuing plating the tops of the studs while the other starts setting the plate. Your partner will get several sticks of plate tacked in place out in front of you, then drop back and help you set the plate. As you catch up to the unplated studs, your partner will again get more studs plated out in front of you. On small jobs, it's common to just plate the entire parapet wall, then go back and set it to the 3-inch chalk line.

Step 3: Setting the Plate Starting at the corner, clamp off four or five studs to the plate. There's no need to pull a layout; the studs are set and can't be moved now. At each stud, measure down from the top of the plate to the chalk line, and use your axe to tap the plate up or down until you get the plate 3 inches above the chalk line. Set all the studs you've clamped first, then screw them off.

Once you've got a full stick of plate set to the chalk line, work your way back to the corner, clamping off four or five studs on the outside of the wall. Then use your 2-foot or torpedo level to set the outside of the plate at each stud, as shown in Figure 8–17. After you get the plate leveled at each stud, you'll again screw the plate off with lathe screws. Screwing off the outside of the plate will be a little awkward, though. To make it easier, turn the screw gun upside down in the palm of your hand and work the trigger with your thumb.

As you get within a few studs of the outside corner stud that you stood up earlier, clamp the corner stud and back out the tack screw holding it in place. That allows you to

Figure 8–16. Here we've slid the first stick of top plate for the parapet wall studs down on the studs, with the end just shy of the studs in the intersecting wall.

Figure 8–17. Using a 2-foot level to set the top plate.

set the plate. Now use your 4-foot level to level the plate out to the corner studs. Once the plate is set, straightedge across the outside face of the studs in the intersecting wall to set the corner stud in line with the rest of the wall studs. Then retack the outside corner stud to the top plate with one lathe screw. You'll add a couple more screws to the corner stud after the intersecting plate is set.

Step 4: Outside Corners Back in step 2, we covered your first move in forming the corner — running the first stick of plate for the two intersecting walls to within $1/4$ inch of the actual corner point. Use the same method when you're setting the intersecting stick of corner plate.

The first stick of plate for the intersecting wall will overlap the plate already in place. To do this, cut off the inside leg of the plate. Cut it back as far as the plate is wide, plus $1/4$ inch for play. If you're using 8-inch plate, cut the inside leg of the plate back $8^{1/4}$ inches from the end. Then slide the stick of plate down onto the studs with the end $1/4$ inch short of the actual outside corner point (Figure 8–18). Before you get too far down the wall, stand up your corner stud and screw it off.

Continue around the wall, following these methods, until you're back at your starting point. And you're finished with this wall.

Figure 8–18. The top plate of the intersecting parapet wall clamped in place to the first stick of plate.

Figure 8–19. This is a good example of the metal stud framing filling in the red iron superstructure.

Parapet Wall Framed to the Red Iron

As always, begin by stocking all your materials and equipment up on the roof. Then use this method to frame the parapet wall when the structural studs of the exterior wall *don't* run up past the roof line. Here the parapet walls, which can be pretty elaborate, are already formed by the red iron. You use the studs to fill in the voids, giving the plywood and gyp board materials something to screw to. See Figure 8–19. On this type of wall, plumb and straight don't really matter. Your framing will simply follow the edge of the red iron. As long as your studs follow layout, you're good to go.

Diagonal Cross Bracing

Before we finish this chapter on structural studs, let's look at diagonally cross bracing a structural metal stud wall. That's the last order of business to complete a structural wall. The cross bracing material is 2-inch-wide 18 gauge strapping that's delivered to the job site in large rolls. Cross bracing added to a wall prevents it from *racking* (being pushed out of plumb by wind shear or seismic forces).

Cut the cross bracing to length on the chop saw and run it diagonally from corner to corner on either the interior or exterior of the wall. Then weld or screw it to each wall stud on top and bottom (Figure 8–20). On long walls, a series of cross braces are fastened to the studs at approximately 45-degree angles. Work the strapping from the top corner diagonally to the bottom corner, pulling as much slack out of the strapping as possible. Clamp it to the studs, eyeballing it straight from corner to corner. Once the cross bracing is welded in place, it's common to shim out the studs to prevent a giant X-shaped bump in the wall.

That's it for structural walls. In the next chapter we'll look at fire-rated walls and ceilings.

Figure 8–20. You can see the X-bracing running from corner to corner on the back side of this 6-inch wall.

Chapter 9

Fire-Rated Walls and Ceilings

In this chapter we'll cover some of the methods you'll use to frame and rock the common fire-rated walls and ceilings you'll face as a commercial metal stud framer. Most of these methods use materials familiar to you from other framing methods we've already covered. But there's one new framing method we'll cover here: the shaft liner system. This will introduce you to an entire new line of materials. The plate, studs and coreboard are all specifically designed for one job and one job only: stopping the spread of fire.

Make no mistake about it, the engineers, architects, and job site superintendents take this work very seriously. Cutting corners on fire-rated walls is a risky proposition at best. Can you imagine rocking the electrical outlet boxes inside the walls? You may think I'm pulling your leg, but it's common to see a couple of guys doing nothing but laminating little pieces of drywall to the electrical boxes with drywall mud, and taping them for 8 hours a day. This is known as *five siding* the boxes. Most foremen have laborers doing this work. It's necessary when the electrical boxes in two separate rooms are put in one stud cavity, to prevent fire from traveling through the electrical boxes from room to room.

You'll also run into some other new materials that are designed to slow or stop the spread of fire. There's special fire caulking, fire-rated insulation and, of course, the drywall will all be fire-rated. I think I've made my point about the importance of fire-rated walls. Now let's get to work, beginning with our first topic, shaft liner systems.

Shaft Liner Systems

The shaft liner (or shaft wall) system is designed to control the progress of fire up through shafts and other large openings in the concrete deck between the floors of a building, as well as area separation walls. This system is commonly used around elevator shafts and openings in the concrete that large HVAC trunk lines pass through. If a fire does break out in the building and reaches the openings in the deck, it'll be contained in the shaft liner. To do this, the coreboard used in the shaft liner system is installed *inside* the framing, in the back of the specially-designed studs that have grooves or tabs to hold the coreboard tight inside the framing. Besides the shaft wall method we'll cover here, shaft liner soffits and ceilings are also framed with this system. Once you understand how the framing and the coreboard work together, there are no limits to what you can build with this system.

Before we go any further, let's take a look at these new materials.

Plate

The shaft liner plate (*J-runner*, which is also called *hi-low plate*), has a 2-inch back leg and a 1-inch front leg to allow easy installation of the coreboard into the framing. See Figure 9–1. The hi-low plate comes in 10-foot lengths and several widths. They're all made of light gauge material that you can cut with your snips. Hi-low plate is used as both top and bottom plate, with the tall leg always at the inside of the shaft and the short leg out. You'll also use the hi-low plate to header around obstacles, frame around elevator doors, and build outside corners for the shaft walls (when outside corner studs aren't available). Shaft walls are sometimes framed as expansion walls, using deep leg (deflection) plate for the top plate of the wall.

Studs

The shaft liner studs come in two styles, with either tabs (H studs) or grooves (C-H studs). Both are available in several widths and lengths (Figure 9–2). The C-H studs, shown in Figure 9–2, are the most common. The tabs or grooves in the studs hold the coreboard tightly in place inside the wall. Always set the grooves or tabs into the back (or inside) of the shaft wall to hold the coreboard close to the edge of the shaft. In a pinch, you can cut either type of shaft liner stud with snips, but it's more efficient to use a chop saw. There's also a 2-inch H-stud used to hold two layers of coreboard inside the framing.

Figure 9–1. This is a stick of $3^{5/8}$-inch shaft liner plate. The shorter leg of the plate always goes to the outside of the shaft. You shoot down the plate just like any other.

Figure 9–2. On the left, you can see a typical $3^{5/8}$-inch C-H style shaft liner stud. On the right is a typical $3^{5/8}$-inch H style shaft liner stud.

Coreboard

Coreboard is 2-foot wide, 1-inch thick sheets of drywall with a one-hour fire rating. There's no need to pull a stud layout on a shaft wall. The coreboard itself will set the layout for the studs. The edges of the coreboard are tapered to make it easy to slide the sheet into the grooves or tabs of the studs.

You can cut the coreboard with your utility knife, like common 5/8- and 1/2-inch drywall, but you'll have to score the cuts three or four times to get it to snap. Even after you've scored it several times, the coreboard will be tough to snap. Whether you're ripping the sheet down to width or cutting it to length, you'll find it's easier to snap the coreboard by slamming your cut over the edge of the stack, or a scrap piece of coreboard, as shown in Figure 9–3. Then run your knife blade down through the cut to make the back cut, and snap the sheet. Use your router to make intricate cuts. Make these cuts slowly or you'll break the bit.

While the whole idea of the shaft liner system is to completely seal up a shaft, there are often pipe and conduit running close to the ductwork that you're building the shaft wall around. That means you'll have to cut out penetrations in the coreboard and split the sheet across the holes, just like topping out a standard drywall wall (which we'll cover in Chapter 13). You'll also run into obstacles that you'll have to header around, and situations where the shaft liner must branch out to form a soffit when ductwork branches off the main trunk line. These complications are just a typical day on the job! You have to be prepared to face them all.

For the sample job in this chapter, we'll build a simple shaft liner to encase an elevator shaft that runs between two levels of the building. As always, your first step will be to stock all your materials, equipment and scaffolding. When building a shaft liner, especially a tall one, you'll need a scaffold built with planks at several levels. The height of the levels will depend on the length of coreboard you're working with. That lets you build the 24-inch-wide sections of the shaft wall by working two or three rows at a time from the floor to the deck. With the multi-level scaffold, you can stand up a sheet of coreboard and a stud, then climb up to the next level and start on the next sheet without mov-

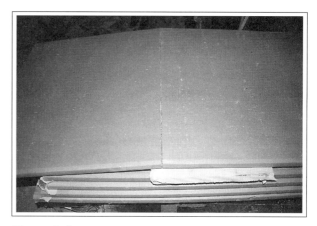

Figure 9–3. With the sheet of coreboard scored, slide a scrap piece of coreboard right under the cut line and then snap it over the scrap.

ing the scaffold planks. Let's go through the process, step by step.

Step 1: Bottom Plate Get the floor swept up around your work area and resnap the layout lines if needed. If the shaft wall hasn't been laid out in advance, there are a couple of things you need to consider before you lay it out yourself.

- One or more sides of the shaft liner may tie into other walls not yet framed. In that case, the shaft wall will follow this wall line, and you must allow for any additional layers of rock on the shaft wall. It seldom has fewer than two layers.

- The shaft liner can also frame up separate from the surrounding walls. Then you'll just make sure the shaft wall jibes with the other walls in the area, and sits square.

- In nearly all shaft wall installations, space is a big issue. You'll want the back of the plate as close to the cutout in the concrete as possible — without letting it hang over.

For this example, we'll assume the shaft liner is already laid out. You'll begin by giving your partner the measurements for the lengths of plate. When it's cut, shoot down the bottom plate all the way around the shaft wall (Figure 9–4), and you're ready to move on.

Step 2: Top Plate From here on out you'll be working up on the scaffold while your partner cuts your material and hands it up. He'll also be building the corners, which we'll cover shortly. You'll work your way around the shaft wall, plumbing, plating and standing one complete wall at a time. Make sure you've got plenty of shot and pins, lathe screws and 1⁵⁄₈- or 2-inch drywall screws on your scaffold, along with your screw gun and shotgun.

Beginning on a long side of the shaft liner, plumb up the wall line and put a 90-degree plumb mark at each end of the wall. That 90-degree plumb mark will save you a lot of time. Use it to establish the intersecting end wall lines and to set the outside corner points of the side you're working on. On many short walls, you can skip snapping a chalk line and just set the plate to the plumb marks. That method will work for this wall.

Your partner will have your plate cut to length and notched for the overlap at the corners. The measurements for the top plate should be the same as the bottom plate. If they're not, something's wrong. Give your partner the first couple of stud lengths, then set the plate to the plumb marks, lining up one end just short of the actual corner point. Again, the deep leg of the hi-low plate will go to the back (or inside) of the shaft wall. Now tack up the plate with a pin close to each plumb mark, then shoot off the rest of the plate (except the corners), spacing the pins 24 inches apart. Look at Figure 9–5.

There's one more wrinkle to keep in mind. You'll often be required to frame shaft liner systems using deflection plate at the top plate. This won't change the procedure except that you'll cut the studs, coreboard and corners 1 inch short of the deck.

By the time you get your first stick of top plate shot up, your partner will have your first stud and sheet of coreboard cut to length. Since the corner is the first thing you'll stand up, let's take a break away from our wall and look at building the corners.

Step 3: The Corners The corners, built out of hi-low plate, are fairly simple to put together. The most important thing to remember is that the tall leg of both pieces of plate that form the corner go at the back of the wall. Set the bottom of the first stick of plate to the tall leg of the

Figure 9–4. Here the hi-low plate is shot down right along the edge of the elevator shaft pit. As the front wall of the elevator shaft is plated, it will follow the same plate line as the 3⁵⁄₈-inch wall that's intersecting.

Figure 9–5. Looking up at the top plate from inside the shaft, you can see it's shot up just like in any other wall. There's a lot of steel in this shot, but you can still see the deep leg of the hi-low plate set to the inside of the elevator shaft.

second stick of plate, as in Figure 9–6. Form the outside corner point with the short leg and bottom of one stick of plate. Flush up the bottom of the first stick and the short leg of the second stick of plate (which is also part of the outside corner) and clamp them together at each end. Then screw together the two pieces of plate with S-12s spaced about 1 foot apart and staggered from side to side.

Once the corner is built, it's pretty hard to cut it to length. So if you're building a corner less than 10 feet tall, cut the plate to length *before* you put it together. For corners over 10 feet tall, you'll splice the plate together just like you splice the top and bottom plates of a wall. When you're building a tall corner like this, cut 24 inches off of one of the first two sticks of plate to stagger the joints. Staggering the joints stiffens up the corner and makes it a lot easier to stand up.

A corner is the first section of the shaft liner you'll stand up, so as soon as you get a couple of corners built, stand up the first one in the plate. As you work your way around the shaft, your partner should find the time to build the rest of the corners before you get to them.

Doesn't it seem like there ought to be a better way? There is! You can see a prefab shaft liner corner (E-stud) in Figure 9–7. When they're available, use the E-studs to build the corners, following the same procedure we covered here. Put the groove of the E-stud to the back of the shaft wall. With either type of corner, you're ready to stand up the wall as soon as the first corner is standing and screwed off in the top and bottom plate.

Step 4: First a Sheet and Then a Stud
Stand a sheet of coreboard, then a stud, then another sheet and another stud, and so on. That's how this framing system works. Measure the length of the coreboard from the floor to the short leg of the top plate, then allow an extra $1/8$ inch to make it easy to stand the sheet. Measure the studs from floor to deck, and allow $1/4$ inch (except for deflection walls).

With the first sheet and stud cut to length, your partner will stand the sheet of coreboard in the bottom plate about 3 inches away from the corner. Then you'll lean the top of

Figure 9–6. Setting the tall leg of one stick of hi-low plate to the bottom of a second stick of hi-low plate. Be sure the two plates are flush, to form a smooth outside corner.

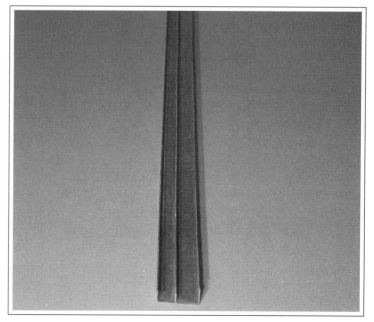

Figure 9–7. Here you can see an E-stud which is used to form shaft liner corners. Stand and set an E-stud corner following the same procedures as a corner with hi-low plate.

Commercial Metal Stud Framing

the sheet back into the top plate, and slide the sheet all the way into the corner. Remember, the coreboard will set the layout for the wall. The rest of the wall will follow this first sheet — so make sure it's plumb. If the sheet needs to be adjusted, run a screw in the bottom corner of the sheet on the side it needs to be tilted toward, to act as a hinge or pivot point. Tilt the sheet to the plumb position and screw off the sheet at both the top and bottom plate. On tall walls, screwing off the bottom plate with a few screws will hold the sheet plumb. Don't screw off the sheet too close to the edge of the sheet that the stud will slide onto. That would make it hard to get the stud all the way on the edge of the sheet.

Now you're ready for the stud. Have your partner stand the stud in the plate and slide it onto the edge of the sheet. This is where the procedure differs, depending on which kind of stud you're using. The tabbed H-studs slide onto the edge of the sheets pretty easily. The grooved C-H studs fit tighter and you'll often have to tap them all the way on to the coreboard edges with your axe. When you do, set a scrap piece of coreboard or a chunk of wood against the edge and tap on that. There are slits inside the grooves of the studs which allow you to see when the stud is all the way on the edge, and you'll soon learn to feel when the stud is tight. See Figure 9–8.

Once you get the stud in place, screw if off in both the front and back of the top and bottom plates with S-12s. With the stud screwed off in the plate, run some $1^{5}/_{8}$ or 2-inch drywall screws through the edge of the sheet into the back flange of the stud. Space the screws approximately 24 inches apart.

Next, get the corner screwed off. Hold the corner (E-stud or hi-low plate) tight to the factory edge of the sheet against the outside corner and run your screws through the coreboard about 1 inch in from the edge. This will prevent the edge of the sheet from splitting out. If the screws are much farther back, they'll be in the way when you slide the coreboard in the opposite side of the corner. If there are any screws in the way later, grab them with your lineman's pliers and bend them over, snapping them off.

From here your wall should frame on up smooth and easy. Stand your next sheet of coreboard in the plate and work it into the groove of the stud. When the sheet's completely in the groove, run some drywall screws through the edge of the sheet into the flange on the back of the stud. Then do another stud and another sheet of coreboard on down the wall to the next corner (Figure 9–9).

Step 5: Turning the Corner When you get to the corner sheet, measure its width from the inside of the stud's groove to the corner point at

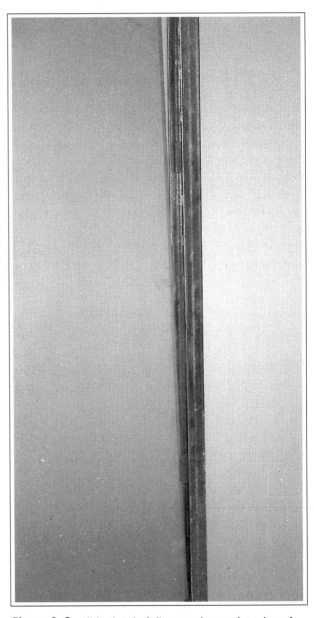

Figure 9–8. Slide the shaft liner stud onto the edge of the sheet of coreboard once the sheet is stood up. As you can see, the clips inside the studs are slipped tightly onto the edge of the sheet.

the top of the wall. Your partner will do the same at the bottom plate. Allow $1/2$ inch off the width so you have plenty of room to set the corner. When the corner sheet is cut, stand it in the wall, slide it into the groove of the stud, and screw it off. Keep the screws in the top and bottom at least 6 inches away from the outside edge of the sheet so you can slide the corner in place.

Snip the front leg of the top plate and fold it out so you can install the corner. Stand the corner up and slide it inside the bottom plate, with the tall leg of the corner slipped between the plate and the coreboard. Then stand the corner into the top plate and fold the leg back in. Next, set your corner to the actual corner point and screw it off at the top and bottom plate.

The first corner we framed up to begin this wall was set to the factory edge of the first sheet of coreboard. For this corner, use a common stud or rip of drywall as a straightedge to set the corner as it's screwed off to the coreboard. On a tall shaft liner (over 10 to 15 feet) where the shaft wall will be part of the finished walls, set the corners with a straightedge and a 4-foot level. If the ceiling is also high, pull a dryline and set the corner to it.

Have your partner turn the corner with your scaffold, then hand you up the top plate for this side of the wall. Set the plate aside for a minute and drop a plumb bob at the far corner, marking both the wall line and a plumb mark 90-degrees to it. Use the 90-degree plumb mark to establish the intersecting wall line (which is also the outside corner point), and you're ready to shoot up the top plate. Slide the top plate up over your corner so the top plates of the two intersecting walls overlap, just as with all the corners. Then adjust the ends of the plate to just shy of the actual corner points. Set it to the wall line plumb marks and shoot it in place. Now you're ready to stand the first sheet and stud of this wall. Follow these steps on around the wall to completion and you're ready for your next project.

Tall Shaft Liner Variations

Shaft liners are sometimes much taller than the length of either the coreboard or studs, so they'll both have to be stacked. The process is similar to framing a regular shaft liner. The key point to keep in mind here is that all the joints of

Figure 9-9. Several coreboard panels and shaft liner studs that have been stood up and screwed off.

both the coreboard and studs must be staggered. Another issue to consider is manpower. If a job looks like it'll take more help to get it done efficiently or safely, talk to your foreman. He'll help you decide if you need the extra help, and get it for you if you do.

We've already covered building corners to any length, so here we'll just look at how to stand the coreboard and studs. For this example, we'll work up each section of the wall individually, but it's common to work up two or even three sections at a time. Experience will teach you what works best for you in any given situation. For your first couple of shaft walls, work up one section at a time until you get a feel for it.

The first section of coreboard will all be full sheets to the deck, with the top sheet cut to fit. Setting the factory ends of the sheets together at the butt joints will help maintain plumb throughout the section. Cut the first stud short of the joint in the coreboard, staggering the joints by no less than 24 inches. Once the first sheet is up, plumbed and tacked, stand the first stud with it. Then stack up the next sheet, and the next stud, as shown in Figure 9–10.

Next, cut 16 inches off the first sheet of the next section, and cut the stud 24 inches shorter than the sheet of coreboard. Install the remaining sheets of coreboard and studs in this row at full length, again cutting the top sheet to the deck. The cuts you've made to the first sheet and stud in this row will automatically stagger the rest of the joints.

For the third row, again, use full sheets of coreboard, and cut the first stud 24 inches short of the joint in the coreboard. This pattern will continue down the wall, with every other section of the shaft liner cut short.

Keep in mind that the prints or the local fire codes may call out for more than a 16-inch stagger in the joints of the coreboard. And you may be required to *back out* or *back block* the joints. When your foreman lines you out on the project, ask him for these details if he doesn't mention them. He won't think less of you for it.

Backing Out and Back Blocking the Coreboard Joints

Depending on the local fire codes and/or how the shaft liner was engineered, you may have to seal the joints in the coreboard. This will maintain the fire rating at the joint, which might otherwise be a weak spot. There are a couple of different ways to do this. Which method you'll use depends on several factors, including the type of shaft liner you're building.

On shaft liner walls like the ones we just covered, the joints are commonly backed out with a scrap piece of shaft liner stud. Once you've stood up and screwed off the first sheet of coreboard, cut off a 22-inch piece of shaft wall stud and slide it down onto the top of the sheet. Center the stud so its ends are inside the edge of the sheet about 1 inch on both sides. As you stack the next sheet in this section, slide the bottom of it down into the groove of the 22-inch stud as you set it in place. See Figure 9–11.

Another common method of sealing the joints in both shaft liner walls and ceilings is *back blocking*. It's simply a 12-inch piece of coreboard 24 inches wide, coated with all-purpose drywall mud on one side. After the two sheets that form the joint are in place, studded up and screwed off, reach around to the inside of the shaft and set the mudded side to the joint. Center the back blocking on the joint. Then use laminating drywall screws (also called *rock to rock* screws) to screw through the back blocking into the sheets of coreboard. Use four screws on each side of the joint (Figure 9–12).

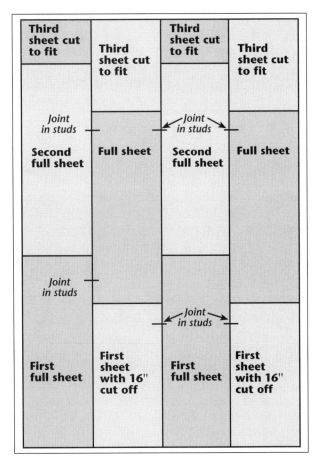

Figure 9–10. Stack the sheets of coreboard and studs as the shaft liner is framed up. Stagger the joints in the studs and the coreboard to make the wall solid.

Fire-Rated Walls

As a metal stud framer, it's inevitable that you'll be called on to hang some drywall. In Chapter 13, we'll cover most of the basic mechanics of hanging drywall. Here we'll focus only on the

Chapter 9: Fire-Rated Walls and Ceilings   137

technical differences that give a wall its fire rating. One of the foremost concerns is that all the joints are staggered between the different layers of drywall. This includes all butt joints, seams and corners of the drywall, whether it's vertical or horizontal.

Corners

A properly-rocked corner is important on any wall, and even more so on a fire-rated wall. To maintain the fire rating of a corner, overlap the rock from side to side of the corner. You'll always be covering the cut (or factory) edge of the sheet that's hung on the opposite side of the corner. As you rock the first side of the corner, cut the drywall flush with the outside edge of the stud. Then cut the rock on the opposite side of the corner flush with the rock on the first side, covering the exposed edge. Any additional layers of rock will simply continue this process (Figure 9–13).

It's common to rock the walls on one side of the corner at a time, setting up the corner for each layer of drywall as you go. Cut the first layer flush to the corner, and the second layer $5/8$ inch (or the thickness of the rock) past the corner, as shown in Figure 9–14. This method continues for as many layers as the wall calls out, with the corner of each layer extending an additional $5/8$ inch past the last.

Figure 9–11. The joint in the coreboard is backed out as the shaft liner is framed. First stand the sheet, then the stud, then back out the joint. Finally, stack up the next sheet.

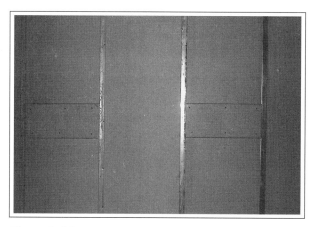

Figure 9–12. Looking at the back of the shaft liner wall from inside the shaft, you can see the joints in the coreboard back blocked with 12-inch pieces of scrap coreboard.

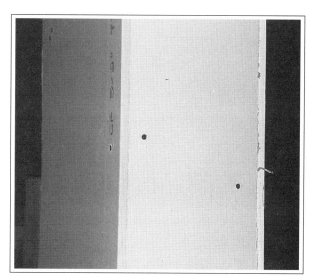

Figure 9–13. The red iron column is wrapped with one layer of $5/8$-inch fire-rated drywall, with the corners of the rock staggered.

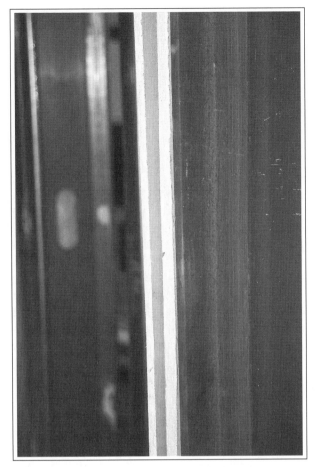

Figure 9–14. Here the drywall is being laminated (glued) directly to the I-beam. The column is being wrapped one side at a time and the individual layers run out 5/8 inch farther at each layer to keep the corners staggered.

Rating a Truss

It's common for a fire wall to be framed up to the bar joist that it runs parallel with. Sometimes the wall won't clear the lip of the bar joist, which will prevent the rock from running up to the deck on one side of the wall. Instead of framing out and around the truss (bar joist), you can often get approval to just wire the drywall right to the truss. Cut the drywall into rips that'll fit inside the truss so that the weight of the rip rests on the lip of the truss. The wire just holds the rock to the truss and doesn't carry any weight.

You'll use 18-gauge tie wire to tie the rock in place to the diagonal bracing of the trusses, with a wire every 48 inches. Once your partner cuts the rips to width, measure lengthwise down the center of the truss to get the measurement for a diagonal brace approximately every 4 feet. Pull these numbers off of what the rip will butt to, allowing 2 inches on either side of the brace. Your partner will then mark the numbers down the center of the sheet, then stab through the rock with a keyhole saw, rounding out holes to about the size of a pencil.

Cut the wires approximately 16 inches long, and bend them in the shape of a large U. Now that you've got everything ready, set the rip on the lip of the truss, and slide it into place. Lean the top of the sheet out and from behind the truss brace, then start the wires through the holes. After you get all the wires started through the sheet, lean it back against the truss and wrap the wires hand tight, as in Figure 9–15.

If you spend many years in this trade, you're sure to come across situations where this method comes in handy. In a lot of chase walls, you'll have to rock the inside of the chase as well as the outside. The last sheet you hang inside the chase would be pretty hard to screw off. This wiring method will do the job.

Fire Ceiling

Some projects will call for a fire ceiling (also called a *fire lid*) to be framed in and rocked before any of the walls are framed up (Figure 9–16). Most fire lids are framed with DWC that's tied directly to the bottoms of the bar joist, using the same tie wire technique we covered back in Chapter 3. You'll also use utility angle around the perimeter of the fire lid just as you would a suspended ceiling, except here it'll go up last, not first.

You'll do all of this work from a scaffold, and you and your partner will each need one. Because the fire lid is one of the first phases of the interior work to be done, you'll usually have a wide-open work area. That makes large 5-foot scaffolds the best choice because they provide a large work area. Because these scaffolds aren't top-heavy, it's common to see carpenters pull themselves along by the bar joist instead of climbing up and down the scaffold. I don't advocate this practice, and only mention it here to warn you *never* to do this on a narrow 2-foot scaffold of any kind, as it'll easily tip over.

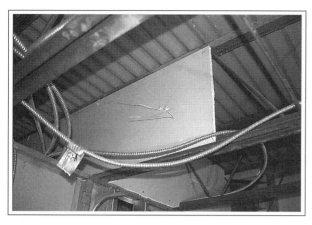

Figure 9–15. This truss is being fire rated by tying the drywall directly to the cross-bracing of the truss.

Figure 9–16. Here you can see a typical fire lid framed to the bottom of the bar joist. The walls will go in after the fire lid is rocked and taped. All the penetrations in this fire lid will be fire taped tightly closed and will often also have to be fire caulked. Once all the remaining walls are completed, a finish ceiling will be suspended from the fire lid.

Now, with your scaffolds set up and all your materials stocked, it's time to get to work.

Step 1: Lay Out the Rows of DWC The rows of DWC will run perpendicular to the bar joist, so pull your layout off the longest end wall that's also perpendicular to the bar joist. Remember that when working with DWC, you'll have to lay out the edge which is $1^{1}/_{4}$ inches off the centers. If you marked the center, the DWC would cover the marks. You and your partner will each take one end of the work area (or start about 50 feet apart on large areas). You'll each pull layout across the two end bar joists as far as you can reach from your scaffold.

Now, pull a chalk line between you and your partner, and starting at the first layout marks, snap a line marking all the bar joist in between. That's what they're doing in Figure 9–17. Getting the end of your chalk line back and forth between you and your partner, who's 50 feet away, could be a problem. Does it surprise you that I've got a trick to suggest? Before you climb up on your scaffolds, tie the ends of both of your chalk boxes together. Now, snap four or five lines, then reel the knot from one partner to the other and snap four or five more.

Using these methods, work your way across the entire work area, laying out the bar joist. Once done, you're ready to tie up the DWC.

Step 2: Tying Up the DWC To begin this step, load both scaffolds up with several bundles of DWC and plenty of tie wire so you don't have to run up and down the scaffold. Align the ends of the scaffolds so that you and your partner can work two rows of DWC at a time. The technique of tying the DWC to the bar joist is the same as the method covered back in Chapter 3, so we won't go through the tying process again.

Hold the first and last stick of DWC in each row back off of the wall about $^{1}/_{8}$ inch, to allow room for the utility angle to slide in place against the walls. After setting the middle of the first stick of DWC to a layout mark, bump it up to the wall and pull it slightly back, leaving the $^{1}/_{8}$-inch gap. Now, set the rest of the stick to the chalk line and tie it to each bar joist, as in Figure 9–18.

Your partner will set the end of his stick of DWC up to yours, overlapping the joint by 6 inches. You can either tie the joints of the DWC with tie wire, or screw them off with panheads. Following this process, tie up all the DWC you can reach without moving your scaffold.

Step 3: Cinch Down the Wires Once you've tied up all the DWC within reach of your scaffold, go back through and cinch down all the wires with your end nippers. Make any needed adjustments to the DWC by tapping it into position to the chalk line with your axe as you cinch it down. Work your way across the ceiling, tying up all the DWC in the first rows, then reposition your scaffold and do the next rows.

Step 4: Utility Angle With all the DWC tied up to the bar joist, load your scaffold with utility angle and grab a screw gun. Now work your way around the perimeter of the ceiling, screwing the utility angle to the wall studs with drywall screws.

Figure 9–17. Snapping a chalk line to mark layout to all the bar joist at once. The ends of two chalk reels are tied together so the carpenters can just reel fresh line back and forth instead of getting up and down from their scaffolds.

Figure 9–18. The DWC being tied to the bar joist. Notice that the side wall has been rocked down below the fire lid but above the finished ceiling height. On the right side of the photo you can also see the DWC running at an angle down under an I-beam that's at a lower elevation than the bar joist. Running the DWC at an angle saves framing the fire lid into the I-beam and then boxing the rest of it in.

Slide the utility angle up behind the DWC, tight to the bottoms, and screw it to the wall. Then use panheads to screw the utility angle to the DWC, as shown in Figure 9–19. The utility angle will simply follow the DWC where the DWC butts to the wall. On walls that are parallel to the rows of DWC, use a straightedge to transfer the ceiling height over to the wall, then snap a chalk line to set the utility angle to. Remember, this is a fire lid, so flat and level aren't issues here. All that matters is that the ceiling seals off the area. A grid ceiling will hang from your fire ceiling later. In rare circumstances, portions of the fire lid (which is often vast) will also double as the finish ceiling. The process, however, will be the same.

It'll be quick and easy to frame the walls to the fire lids. Once the ceiling is rocked and fire taped, you'll have a smooth, clean surface to frame to. Screw the top plate of the walls to the ceiling with drywall screws, screwing through the rock into the DWC. Cross screw the top plate of the walls running parallel to, but between, the rows of DWC, to the drywall just like you tie in a slider to a wall. We'll cover that in Chapter 13.

Let's move on to Chapter 10 and look at a couple of ways to frame up a column.

Figure 9–19. You can see the utility angle screwed off to the wall studs; the ends of the DWC are then screwed off to the utility angle.

Chapter 10

Columns

Nearly all columns are built for one purpose and one purpose only: to hide something. Whether a column is plain or embellished with angles, offsets or crowns, you can bet there's something hidden inside it. Red iron piers, ductwork, water lines, and electrical conduit are commonly enclosed in columns, either because they don't fit into a wall or they missed the wall.

In this chapter we'll cover two common methods used to frame up a column, plus a simple method of framing a cornice (crown) to the top of a column. One of the first things you'll notice is how quick and easy it is to frame the columns with metal studs, no matter which method you use. The step-by-step explanations may seem complicated, but you'll cruise through the job when you put these methods to work. Unless they've been mishandled or stepped on, a metal stud is nearly always perfectly straight. That makes for a high-quality finished job.

Keep in mind that there's no limit to the situations you'll run into. Of course, you also have a wide variety of materials at your disposal. When push comes to shove, use whatever it takes to get the job done. I've built many columns in really tight spots that are solid as a rock, using just drywall and wall molding (90-degree grid ceiling material).

Before we get started, I want to point out one common and very costly mistake you can make when framing columns. Whenever you're framing a column up against a fire wall, make sure you fire tape *all the joints* in the drywall behind the column before they're covered up. It's easy to overlook this. And the mistake might never be caught. But when it is, the time you saved is only a fraction of the time you'll spend on making it right. There's no profit in rework.

Full-Framed Columns

This is the usual method of framing a column that's built around a weight-bearing red iron or concrete pier. But before you start building, there's one important factor to consider. The key to properly framing up a column, using this or any other method, is *laying it out right*. This is particularly important in areas that will have either a grid ceiling or computer floor (or both) installed later. The goal is to build all of the columns in straight, even rows that are square with the grid lines. Unless the columns are laid out in advance by the general contractor, you'll have to pull layout off of both the north/south and east/west reference lines. If for any reason the reference lines haven't been established, check with the foremen of the grid ceiling and computer floor crews so that everyone is in agreement on the reference lines. Better yet, let *them* establish the reference lines, making them responsible for any problems down the road.

There's one more thing worth mentioning. If the area you're working in will have a computer floor installed, all the elevations for your framing will be figured from the computer floor height. Most computer floors are installed about 2 feet up off the slab. Your outfit's estimator should have ordered material long enough to accommodate that extra length to the slab.

Step 1: Bottom Plate Once you've established at least one side of the column frame line from the reference lines, use your framing square to lay out the rest. Next, shoot down the bottom plate, overlapping the corners just as you would on the outside corners of a typical wall. In Figure 10-1 you see the bottom plate of the column shot down. Once it's shot down, use the bottom plate as a jig to build the top plate as an exact duplicate. You can accomplish the next two steps much faster by splitting up, with one partner putting the top plate together as the other works the lookouts.

Step 2: The Lookouts Cut the lookouts from scrap stud that's at least as long as the width of the column. With the lookouts cut to length, set them across the bottom plate just as you'll shoot them to the red iron. Slide the lookouts up against the red iron and adjust them by eye so the ends are even on the plate on each side. Then mark the red iron to the hard side of each lookout (Figure 10-2).

The height of the lookouts is determined by the stud length, which we know is no less than 4 inches over the ceiling height. In this example, we're using 9-foot studs, so the lookout height will be $3/8$ inch higher than that. Mark the lookout height to the red iron and level it around to both sides that the lookouts will be shot to. Then just set the bottom edge of the lookout to the leveled line, and line up the mark on the hard side with the same side of the red iron. Finally, shoot it in place with two pins, as shown in Figure 10-3.

Figure 10-1. When the column is laid out squarely with the grid lines or other wall lines in the area, shoot down the bottom plate just like any other bottom plate.

Step 3: The Top Plate Cut the pieces of top plate to length using the bottom plate as a guide. Slide the top plate into the bottom plate on two opposite sides of the red iron, using the bottom plate as a jig. Cut the inside legs of the other two sides of the top plate and fold them

Figure 10–2. After it's shot down, use the bottom plate as a guide to determine the placement of the lookouts. Center the lookouts on the bottom plate and mark the edges of the red iron (or concrete) column. Use these marks to keep the lookouts in the same position when they're shot in place.

out. Don't bother to use your tape. Use the other pieces of plate as a guide and eyeball the cuts. Now slide these two pieces in the bottom plate and screw off the corners with four panheads each, as shown in Figure 10–4.

Plumb the plate line up to the lookouts on one side of the column, also marking a 90-degree plumb mark right to the top plate (which is still set in the bottom plate). You'll need to plumb up a 90-degree plumb mark to one of the lookouts, again marking the 90-degree point on the top plate. You'll use these four points to set the top plate.

Step 4: Putting It All Together Set the top plate to the plumb points on the lookouts, first setting it to the wall line you plumbed up, then to the 90-degree plumb points for the opposite direction. Once it's set, clamp the plate to the lookouts, then tack it off, as shown in Figure 10–5. Turn the studs into position in the corners, adding any needed layout studs. Then screw them off in both the top and bottom plate. Use your level to double-check the edge of each corner stud for plumb. Once you're satisfied that the column is plumb all the way around, screw the top plate to the lookouts with two screws at each point. Figure 10–6 shows the finished column.

Figure 10–3. The marks made earlier on the lookouts are lined up with the same edges of the red iron column and shot in place.

Figure 10–4. Using the bottom plate as a jig for the top plate eliminates a lot of time spent plumbing up. It also ensures the top plate will be the exact same shape as the bottom plate. This technique has many uses.

Figure 10–5. With the plate line plumbed up to the lookouts, clamp the top plate in place. Make any needed adjustments and screw off the top plate to the lookouts with two framing screws at each corner point.

Figure 10–6. With the studs stuffed up and screwed off in the corners and on layout, this column's done.

This basic method will take care of nearly all column framing situations. Of course, when the columns are framed to the deck, the job is a lot simpler. Next we'll take a look at some variations to the basic method.

Columns with Corner Guards

Whether they're framed to the deck or to lookouts, many of the columns you build will require additional studs in the corners. This is most likely to occur in high traffic areas. After the column is rocked, the corner guards are fastened to these additional studs. The procedure of framing the column is the same. The difference is that each corner will get two studs: the standard corner stud plus a second stud, usually 3 inches back from the actual corner point.

Framing a Cornice to the Column

The cornice itself is a stair-step crown that tops off the column. The cornices are often added to columns that are built in areas with stair-step soffits. In that case, the elevation of the cornice will coincide with the elevations of the soffit. Frame the cornice to the column after it's completed, using a second set of lookouts screwed to the column. Then you'll plate the lookouts and add the studs to form the cornice.

The width of material you use to frame the cornice depends on the difference between the bottom of the cornice and the ceiling height. For instance, if there's a 4-inch difference, you could use 8-inch material, which would get the cornice $4^{5}/_{8}$ inches above ceiling height. But keep in mind that the cornice doesn't have to be 4 inches above ceiling height, so 6-inch material would also do the job.

Step 1: The Lookouts You'll determine the length of the lookouts by adding the overhang of both sides to the size of the column. The ends of the lookouts must be cut square, so either cut them on a chop saw or use your speed square if you're cutting the lookouts with snips. Again, work hard to develop your eye so you'll be able to make square cuts by eye.

Step 2: Establish the Lookouts' Elevation If there's other elevation work, like a soffit in the area, the cornice will likely match this already-established elevation. Use a water level or pocket laser to transfer the elevation from the soffit to the column studs. Mark the two studs on each side of the column that the lookouts will screwed into, then establish the lookout elevation from there.

Step 3: Set the Lookouts Measure back from one end of each lookout and put a mark on the bottom to set the overhang. Now set the lookouts to the level line on the column studs, lining up the mark with the outside corner point of a stud. Then clamp the lookout in place. Before you screw off the lookouts, check the corner studs to be sure they're maintaining the proper width. Once everything's set where it needs to be, screw the lookouts to the studs with two tek screws in each stud, as shown in Figure 10–7.

Next, cut two pieces of stud (or plate) to fit in between the lookouts on the two sides of the column opposite the lookouts. Now set the studs to the column, line up the bottom of the studs with the lookouts, and screw them off to the corner studs. See Figure 10–8.

Step 4: Plate the Lookouts The plate will overhang the column by 4 inches, just like the lookouts, so you can use the lookout length for the plate. Have your partner cut the plate

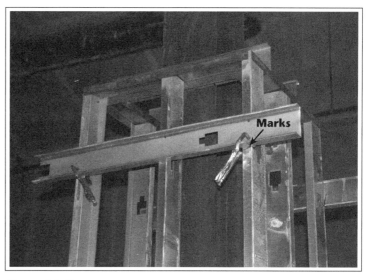

Figure 10–7. The lookout studs set and screwed off to the column studs with two screws in each stud. If you look closely you can see both the overhang and elevation marks on the framing members.

Figure 10–8. Once the lookouts are in place, these short pieces of stud on the two opposite sides of the column will form the inside corners.

to length, again making sure that the cuts are square. Then measure back from one end and put a mark at $3^{3}/_{4}$ inches. Slide the plate onto the ends of the lookouts and line up the $3^{3}/_{4}$-inch mark with the hard side of one of the lookouts. See Figure 10–9. This will automatically set both ends of the pieces of plate.

Now adjust the bottom of the plate so that it's exactly 4 inches off the column studs at each lookout, clamping them in place as they're set. As you set the plate to the opposite ends of the lookouts, watch that it's 4 inches off of the column, and check the outside plate-to-plate number to maintain the cornice width. Next, screw off the plate to the bottom of the lookouts with a single panhead at each lookout. Clamp the plate to the top of the lookouts. Using your 2-foot or torpedo level to plumb the plate, screw it off to the tops of the lookouts, as shown in Figure 10–10.

Step 5: Stud Up the Plate Like the lookouts and the plate, the cornice studs will also be cut 30¼ inches long. Once cut, slide the studs into the ends of the plate on each side of the column, with the hard side of the studs out. Set the bottom of the studs 4 inches off the column on each side, again checking the outside stud-to-stud measurement to be sure it's even. When it is, screw it off in the plate. Plumb the hard side of the studs in the plate and screw them off to the tops of the plate. You can see the completed cornice in Figure 10–11. In Figure 10–12, the column and cornice are shown completed and ready for the rockers.

Speed-Framed Columns

Speed framing is a great trick that will save you a lot of time. Speed framing uses the drywall itself to frame up the column. That means the rips will set the outside corners, so they must be cut straight and the edges cleaned. This method does have limits, though. You can only speed frame a column that has no more than three sides, to an already framed and rocked wall. If the walls are framed up, and items like pipes, conduit or red iron missed the wall line, this is a quick, easy solution to the problem.

Keep in mind that when you use this method, you're committed to the already-built wall,

Figure 10–9. Slide the plate onto the ends of the lookouts and set the end-o (overhang) mark to the hard side of the lookout. Set the cornice width off of the column studs as well as the out-to-out overall width of the plate.

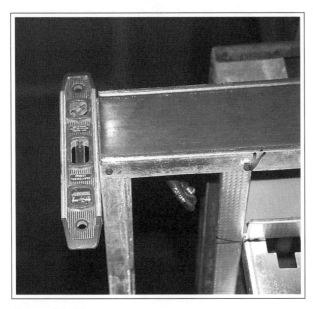

Figure 10–10. With the width of the cornice set at the bottom of the stud, a torpedo level is used to plumb the hard side of the stud.

Figure 10–11. The studs, added to the plate and set for width, complete this very simple cornice.

Figure 10–12. A full-height view of the column and cornice. Once the column and surrounding walls are rocked, finished and painted, the grid ceiling will set off the coinciding elevation of the cornice and the double windows in the area.

Figure 10–13. Alternate the screws up and down to tie the column studs directly to the drywall of the walls that are already framed and rocked.

which may or may not be exactly plumb. As long as the wall isn't too far out of plumb, don't worry about it. Just make the column follow the wall, or you'll have a visible difference between the wall and column. For this example, we'll build a two-sided column in a corner — but you build a three-sided column the same way. It just gets a third rip for the front of the column.

Step 1: Bottom Plate You'll do the layout work with your framing and combination squares, beginning with one of the sides that'll be squared off the back wall. Then simply measure off of the wall, marking the front plate line. Since this column is in a corner, it'll only have two sides. So just square off of the back walls.

As you're laying out the bottom plate line, your partner will cut the pieces of plate to length, cutting them so the corners overlap. Get your plate shot down and you're ready to move on.

Step 2: Tying In the Studs Stand the studs one at a time in the bottom plate and slide them up against the wall. Plumb the studs with your 4-foot level. Tie them into the drywall wall just like you tie in slap studs. Make sure they're solid, because they'll carry the column. If the studs aren't tied in properly, the column will pull loose from the wall. The cross tying method works the best. Whenever possible, screw the column studs to a wall stud. In Figure 10–13, both of the column studs have been properly tied in.

148 Commercial Metal Stud Framing

Step 3: The Rips As you're tying in the studs, your partner will cut the rips to width, making sure they're straight. Once the rips are cut, rasp the cut edges smooth. Cut the rips for the sides first, at a width determined at the bottom plate. Cut the front rip to overlap the raw edges of the side rips.

With the rips cut, stand the first side up tight to the wall and screw it off to the stud and bottom plate, spacing the screws 8 inches apart. See Figure 10–14.

Step 4: Add Front Studs and Rip With the first side rip stood up and screwed off, stand a stud in the outside corner with the hard side of the stud up against the rip, lining up the stud on the edge of the rip. Screw the stud off to the rip with screws spaced 8 inches apart. See Figure 10–15.

Now set the second rip in place and even the edges at the bottom plate. Tack it with one screw. Line up the corner so that the edge is even all the way up and screw it off from top to bottom. Then screw off the outside corner, using the edge of the rips to set the corner, as shown in Figure 10–16. The column's finished.

The column and cornice are finished—and so is this chapter. In the next chapter we'll walk through the steps for framing soffits.

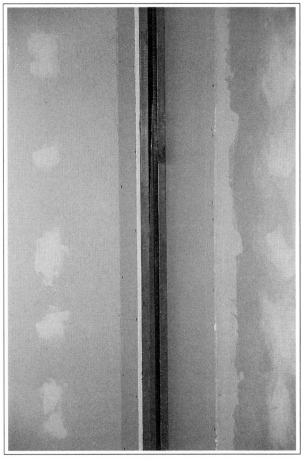

Figure 10–14. The first rip screwed off to one of the studs tied into the wall. The rip is cut straight and rasped clean to form the outside corner. On a three-sided column, set both the side rips first, just like this, then add the front.

Figure 10–15. With the first rip stood up and screwed off, a stud is set to the outside edge of the rip and screwed off. This stud will follow the rip and form the outside corner. On a three-sided column, this step is repeated on both sides of the column.

Figure 10–16. Setting the outside corner of the column to the edge of the rips by leaving a drywall screw hanging out, then pulling it into position with end nippers.

Chapter 11

Soffits

In this chapter, we'll take a look at a few techniques used to build soffits (also referred to as *drops*). This is a framing job that you'll run into on a regular basis. The soffits are often used like columns, to hide exposed ductwork, plumbing lines and electrical conduit. You'll use a soffit to fill in the void between cabinets (casework) and the ceiling, and to house recessed lights above countertops and display cases. Soffits can also hold light troughs for indirect lighting.

There's no limit to the sizes and shapes that soffits come in. They range from a basic square box to the very elaborate, with various stair steps, offsets and angles. We'll cover several methods here, ranging from using a few scrap studs and a couple of rips to framing the soffits with jigs.

You can find soffits in three different sections of the prints; the casework section, the reflective ceiling plan and the floor plan. An elaborate soffit will also have a cut in the specifications section that shows the exact dimensions of any stair steps, or other unique features. On many large projects, you'll have to refer to all three sections of the prints for proper placement and dimensions of the soffits. You'll also need to check the HVAC, electrical and plumbing sections and talk to the foremen of these trades to find out what will run through any large, time-consuming soffit you're building.

You can imagine the lump you'd get in your throat after completing a soffit that took three days to build, only to find out a duct runs through it. Now the soffit has to come apart so the mechanical sub can install it. I've seen it happen to the best of hands. Remember, failing to prepare is preparing to fail.

Keep in mind that my examples are basic soffit methods built under ideal conditions. You won't often find conditions like that. In fact, you'll run into adverse conditions on most every job. But you can vary each of these methods to handle just about any complication that you run into. Having strong skills in soffit work will keep you working longer on a project, and usually at a higher pay rate.

As we walk through each of these examples, we'll be working off of a 2-foot scaffold, which is common for this type of work. You wouldn't want to waste time tearing down and rebuilding scaffolding as you move from room to room. Stilts will also save you a lot of time on soffit work, *if* you know how to walk on them. They can also get you hurt real bad if you don't. It's a skill worth learning.

But here comes the boss. Let's get to work.

Spanning Speed Soffit

This method of framing the drops uses the rips of drywall themselves help to support the soffit, so you don't have to completely frame it in. Because the rip will help support the soffit's weight, there's a limit to how long a soffit you can build. You can't use it for any soffit longer than a sheet of drywall. You'll also be limited by the overall size of the soffit. About the biggest soffit you can build with this method is 24 by 24 inches. For soffits that fit these restrictions, this method will save you both manhours and material. It's great for casework soffits and soffits in walls that will have phones and drinking fountains set inside them. For added strength and rigidity, use ⁵/₈-inch drywall to build the soffit whenever possible.

For the example illustrated here, we'll be building a soffit that spans wall to wall, creating a slot for cabinets to fit into. The finish elevation for this soffit is shown in the reflective ceiling plans at 7 feet ¹/₂ inch, so the 7-foot cabinets will fit neatly inside. The room will later get a grid ceiling at 8 feet.

As we build this soffit between the two side walls of this room, keep in mind that this method will be just as effective when building a soffit that spans a concave portion of a wall. You can also put this method to work when building a soffit between two columns. The key, as I said, is the length and overall size of the soffit.

Step 1: The Elevation   Begin the soffit by establishing a constant elevation with a water level or laser — I use my pocket laser for this work all of the time. Put a bench mark at each corner point. Now check the floor, measuring down from the bench marks to find the high spot. The cabinets will be leveled to the high spot in the floor. From that high spot, measure up and mark the frame line elevation of the soffit. When you allow ⁵/₈ inch for the rock, that's a frame line elevation of 7 feet 1¹/₈ inches. Next, measure the difference between the bench mark and the elevation mark, and use the difference to transfer the frame line up from the rest of the bench marks. Snap out the plate line and you're ready to go.

Step 2: Corner Studs   The soffit will extend 4 inches above ceiling height, so cut the corner studs and the vertical face rip of this soffit to 16 inches. Once cut, line up the studs on the frame lines about ¹/₄ inch above the frame line elevation. Get the studs set and screw them in place with 1¹/₄-inch drywall screws spaced 6 inches apart. When possible, stagger the screws from front to back, with the screws in the front close to the outside edge of the stud to form a strong joint. In Figure 11-1, you can see the corner studs properly set and screwed off.

Step 3: Stud the Back and End Walls   Install the back wall stud first, from wall to wall. Keep in mind that you could also use plate or utility angle. All you need is something to screw the bottom soffit rip to. Set the hard side of the stud to the back wall and line up the bottom of the stud on the frame line you snapped out earlier. Once it's set, screw the stud off to the wall studs with drywall screws. One screw in each wall stud will do the job; just be sure they don't strip out.

Cut the end wall studs to fit between the back wall stud and the vertical face studs on each end of the soffit. Set them to the frame line, and when possible, screw them off to the wall studs. If you can't hit enough wall studs to fasten them securely to the wall (this is common on the ends), tie the studs in to the drywall using the cross-tying method. See Figure 11-2.

Step 4: Span the Vertical Face Studs Your next move is to span the vertical face studs with two studs cut with tabs on each end, one stud at the top and one at the bottom. Clamp the tabs of the bottom stud to the face studs and set the bottom even with the frame line elevation. Then screw off the tabs using two tek screws in each.

Set the second stud to the top of the face studs, eyeball it level, and screw it off. I prefer to use two spanning studs to stiffen up the face of the soffit, but it's common to use just one at the bottom. Figure 11-3 shows what little framework there is completed. It didn't take much longer to put together than it did for you to read these last three steps.

Step 5: Front Rip The cut edges of the front and bottom rips will form the outside corner of the soffit, so the cuts must be straight. It's tempting to use the factory edge of the rips to form the outside corner of the soffit (and you'll see other carpenters doing it). But this generally causes extra work for the tapers. The recessed factory edge buries the corner bead and makes it hard to finish.

As your partner cuts the rips, it saves a lot of time to cut both the vertical face and bottom rips to width and length before snapping apart the sheet. Allow $1/8$ inch on the length of the rips, so you don't have to fight them into place. That helps prevent the ends of the rock from breaking out. Another time-saver is to fold the top rip all the way back over and rasp the cut edges of both rips at the same time. Then snap the rip off the sheet. Your partner will cut the rips while you're screwing the studs to the back and end walls.

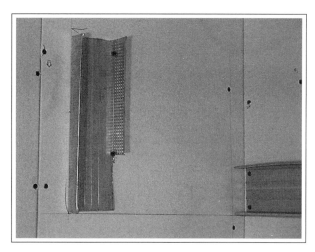

Figure 11-1. The vertical face stud of the soffit tied into the drywall of the wall. The bottom inside corner of the stud is cut out to allow the bottom end stud to run all the way out to the corner.

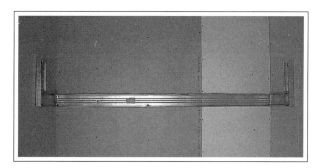

Figure 11-2. The horizontal back and end studs of the speed soffit set to the elevation frame line and tied into the walls.

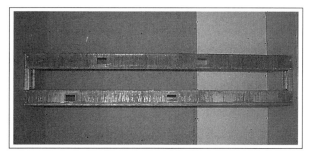

Figure 11-3. The vertical face studs spanned by two $3^{5}/_{8}$-inch studs with tabs cut on each end. This completes what little framework is involved with the speed soffit.

Long rips will take both partners to hang. Set the front rip in place and line up the bottom edge with the edge of the bottom spanning stud. Once set, screw the rip off all the way around, spacing the screws 8 inches apart, as shown in Figure 11-4.

Step 6: Bottom Rip Lift the rip into place and slide it tight to the back wall, splitting any gap evenly between the two ends. Once set, screw off the back and end studs, again spacing the screws 8 inches apart. Just tack up the outside corner for now.

With the bottom rip tacked up, check the outside corner along the vertical face with a straightedge long enough to span the soffit. A stud or a rip with a factory edge will work. See Figure 11-5. Make any needed adjustments to remove any dips or bumps along the edge. When it's set, screw off the bottom of the outside corner.

That completes the soffit — maybe. On long spanning speed soffits like this, the middle along the top edge may feel a little spongy. The grid ceiling will stiffen it up with no problem. But if you (or your boss) want it stiff before the grid goes in, run a little kicker from the back wall over to the top spanning stud in the middle of the soffit. Be sure to check the top of the soffit with a straightedge if you kick it off. This is to prevent the kicker from forming any dips or bumps. Before we move on to the next soffit method, let's take a look at a variation of this one.

Floating End

Not all soffits will span from wall to wall. Many tie into only one end wall, leaving the other end floating. You'll follow the same procedure as the spanning speed soffit method we just covered, except for the floating end. Because most of this method is the same, we'll skip most of the steps and focus on the floating end. While we're putting this technique to work on a spanning type speed soffit, keep in mind that you can use this method in many situations.

With the soffit frame line established and laid out, tie the end stud into the wall, then the one vertical face stud at the end wall. In most cases, neither of these studs will hit on the wall framing, so tie them into the wall using the cross-tie

Figure 11-4. Here you see Julie screwing off the front rip of the soffit.

Figure 11-5. With the bottom rip of the soffit tacked in place, Julie's using a rip of drywall as a straightedge to set the bottom outside corner.

method. Remember, these two studs will carry most of the soffit's weight, so make sure they're secure. Once these two studs are set, screw them off to the wall, as shown in Figure 11-6.

At the (floating) end of the soffit, establish the frame line and plumb it up. Now set the end stud to the plumbed frame line and screw it to the wall (Figure 11-7). The floating end consists of nothing more than a perfectly square cut piece of drywall that's 5/8 inch shy of the overall soffit width. That allows for the rock on the face of the soffit. The length or height of the end is equal to the difference between the frame line elevation and the ceiling height, plus 4 inches. This small piece of rock will set the soffit — and any flaws in it will show in the soffit. Use your T-square or framing square to cut it, and rasp all the cuts smooth.

Once cut, set the piece of rock in place tight to the wall, with the bottom lined up with the frame line elevation. Now screw the rock to the end stud, spacing the screws 6 inches apart. Make sure they don't tear through the paper. Next, cut a stud as long as the end piece of rock is tall. Set it flush with the outside edge and screw it off. Then cut a short piece of stud to fill in the bottom edge and screw it off flush with the bottom edge. The end piece of rock is shown completed in Figure 11-7.

The Spanning Studs

This is a pretty short soffit, so we'll use only one spanning stud at the bottom outside corner. To get the length of the spanning stud, measure the inside length of the soffit along the back wall, minus $2^{3/4}$ inches for the legs of both vertical end studs and $1/4$ inch of play. That gives you the in-between number. The tab at the end wall will be no more than an inch long, and will be pushed tight to the end wall stud so that all the slack is at the floating end of the soffit. Set the spanning stud in place and clamp it off at both ends. Set the elevation at the end wall stud and tack it with one tek screw. Now square the floating end off the wall with your framing square, set the elevation of the spanning stud at this end and screw off the stud with two tek screws. Finally, add a second screw to the other end. See Figure 11-8.

Suspended Speed Soffit

Instead of using tabbed studs and a rip to carry its weight like the spanning soffit, the suspended speed soffit uses a top plate to carry the weight. You can alter this method to work in many situations, but it's most effective when framed to a grid or suspended drywall ceiling. In this case, you use the face rips (if they're cut straight) to set the bottom of the soffit for soffits that are only one or two sheets long. On longer soffits, run the rips wild and set the elevation with a laser. That's one of the key benefits to this method — there's no limit to the soffit's length. That makes it useful for framing drops around the perimeter of a ceiling to form a coffered ceiling. The drop around the perimeter of

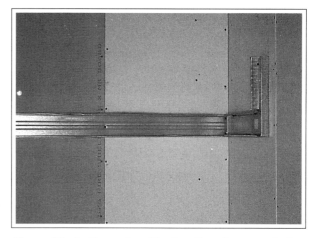

Figure 11-6. The end and back wall studs screwed off to the wall studs where possible and tied into the drywall otherwise.

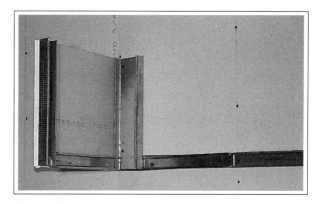

Figure 11-7. A stud tied into the back wall at the end of the soffit supports the floating end. The square-cut piece of drywall (floating end) is then screwed off to the stud. Pieces of stud screwed to the floating end at the front and bottom outside corners complete the floating end.

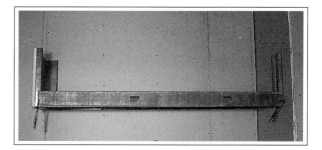

Figure 11-8. The bottom outside corner of the soffit formed by a $3^{5/8}$-inch stud with tabs cut on each end and screwed off to the two end studs.

the ceiling in Figure 11–9 was built using this method.

For our example, we'll build a drop to a suspended drywall ceiling using 1⁵⁄₈-inch framing material and ⁵⁄₈-inch rock. The walls and the ceiling of the area have been rocked and any needed fire taping has been completed. This is a big concern when building soffits. Any time you're building a soffit that ties into any fire-rated walls or ceiling, *make sure it gets fire taped,* even if you have to do it yourself.

Step 1: Get It Plated Screwing off the plate (or stud) close to the flange prevents the weight of the soffit from pulling down on the edge of the plate, which can cause a sag in the soffit line. Since we're building a gravy soffit here, we measure the plate line off the back wall at each end of the ceiling and snap a line, allowing ⁵⁄₈ inch for the rock.

Set the plate to the chalk line and splice it together at the ends, just like the top plate of a wall. Screw the plate to the ceiling with two drywall screws, close to the outside flange of the plate, at each stick of DWC. Watch the screws to make sure they don't strip out. That's a common problem when you screw together two pieces of framing material that are separated by drywall. But it's an easy problem to overcome. Just run the screws in slowly, or use hi-low screws that are double-threaded.

To establish elevation of the back and end wall frame lines, we'll use the existing work, measuring down off the ceiling at each corner point of the soffit. The back and end wall material doesn't have to be plate—you can use stud or utility angle instead. On this soffit we're using studs. Screw the plate to the wall studs around the soffit's perimeter, tying in any ends that don't hit a wall stud. Screw off the vertical face studs in the top plate. Then plumb them with your level as you screw them to a wall stud or tie them into the drywall. In Figure 11–10 you can see the completed framework for the suspended speed soffit.

Your partner will have begun cutting the rips to width, having calculated the width of the rips where the soffit is laid out on the end wall. Cut a few studs to length, about 6 inches short of the soffit's vertical face height. You'll use these studs to splice the joints of the rips together. On

Figure 11–9. This is a suspended speed soffit which is continuous around the entire restaurant.

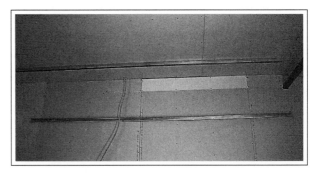

Figure 11–10. You can see the 1⁵⁄₈-inch studs screwed off to the fire lid (establishing the width) and the back wall (establishing the height).

long soffits like the one in Figure 11–9, drop a stud every 4 feet to help carry the weight.

Step 2: The Rips With the rips cut to width and rasped clean, you can start at either end to begin hanging the vertical face rips. Unless you're building a short soffit, this will take both partners. The rips must be held tight to the ceiling evenly all the way across the soffit. If one end is tight and the other drops just a little, it'll put a bump in the bottom outside corner of soffit line. The soffit line is the critical part of the soffit. It's what catches everyone's eye when they look at a soffit, and bumps or dips really stand out.

Once it's set, screw off the rip along the top plate and end stud with screws spaced 8 inches apart. Keep the screws at least ³⁄₄ inch down from the edge of the sheet and make sure they don't tear through the edge of the rip. Figure 11–11 shows the sheet correctly screwed off. The front rip carries the weight of the soffit along the top plate. If the screws tear through, the soffit comes down. Because of this, many carpenters will turn the top plate on edge so they can screw off the rips 2 inches below the top edge of the sheets. I use this technique any time there will be any weight hanging on the soffit—it gives the screws a little more rock to bite into.

With the sheet screwed off to the top plate and end stud, add a break stud to the joint. Turn the stud hard side out and screw it to the front edge of the top plate (Figure 11–12). Get all the front rips hung, and you're ready to move on.

Step 3: The Plate Line You can form the bottom outside corner (plate line) of the soffit using plate, stud or utility angle. Set the first stick of plate to the back side of the rock and line it up with the bottom edge of the rips. Once set, screw off the plate with screws spaced 8 inches apart, keeping the screws at least an inch up from the bottom edge of the rips. See Figure 11–13. Splice the plate together at the joints across the bottom of the tips to the far end and cut the last stick just shy of the end wall stud.

Step 4: The Bottom Rips With the bottom outside corner material screwed to the front rip, we're ready to hang the bottom rips. Begin at the same end of the soffit where you started the front rips. Cut 24 inches off the first rip to stagger the joints between the face and bottom rips. Lift the first rip into place and push it tight to the back wall. Screw off the back and end wall, again spacing the screws 8 inches apart. For now, you can just tack the front of the rip.

Work your way on down the soffit, hanging all the bottom rips. When that's done, you should have about 1/8 inch offset between the front and bottom rips. Use the offset in the rips to set the outside corner, keeping it even all the way down the soffit. Once the outside corner is set and screwed off, sight down the corner to make sure it's straight. Any flaws are easily corrected by backing out a few screws and adjusting the corner. If there's a bump, simply push it back in and rescrew it. Correct a dip by running a screw about half way into the corner and pulling out on it with your pliers. When it's set, rescrew it. Figure 11–14 shows the soffit completed.

If you're ever in doubt about whether the rips are straight or the ceiling is flat, pull a dryline

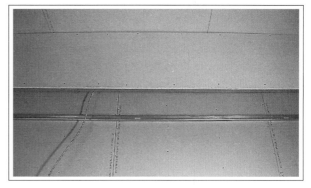

Figure 11–11. The face (front) rip of the soffit screwed off.

Figure 11–12. Determine the break point of the face rip and screw a stud to the ceiling stud, giving the butt joint of the face rips something to break on.

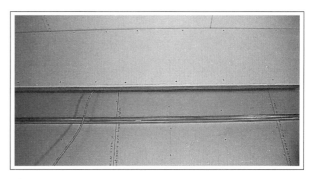

Figure 11–13. The bottom outside corner stud set to the bottom edge of the face rip and screwed off every 8 inches.

and set the corner to it. You'll still save a bunch of time.

A common variation to this method is to run the vertical face rip wild a few inches. Then shoot in the bottom frame line elevation with a laser around the entire soffit and snap a line to the back side of the rips. Set the outside corner material to the chalk line and screw it off from the front. Finally, cut off the waste with a razor knife and rasp it smooth. That's the method we used to build the large soffit in Figure 11–9. If you look at that photo, you can see it's well over 24 inches wide. Once the front rip was set, we framed a small suspended ceiling using DWC from the wall to the front rip. We then added CRC and suspension wires to carry the weight. We completed that soffit in about half the time it would have taken us to fully frame it. If you want to make more money, that's exactly how to do it.

I want to point out one more variation before we move on. Any time you build anything to a grid ceiling, you must allow room for the trim metal (L-metal) that the finishers use. Here's an easy way to do it. Just cut a few pieces of cardboard shim about 6 inches long, and use two of them stacked up at each end of the rips as spacers between the rips and the grid ceiling. J-mold is another trim metal that's often used in situations like this. It slides onto the edge of the rock as it's hung.

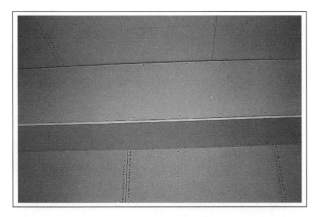

Figure 11–14. The completed suspended speed soffit.

Framed Soffit

Now let's walk through the steps to frame a simple soffit. As with most of the examples we've covered, this soffit will be built under the best of conditions. We'll also branch off from time to time to look at some ways to deal with adverse conditions you might run into.

We're framing a soffit out of $3^{5/8}$-inch material to a flat, clean deck that's fairly level and free of obstacles. Once again, getting set up is the first step. Part of that first step is getting lined out by your foreman and studying the prints. Pay close attention to the prints for HVAC lines, defusers and recessed lights that are common in soffits.

With the material stocked and equipment set up, let's get to work before the boss catches us standin' around talkin'.

Step 1: Lay Out the Soffit The first move is to establish the soffit's plate line elevation around the back and end walls with a water level or level, then snap them out. As with all elevation work, allow for the thickness of the drywall.

Next, plumb up the plate line. This soffit was laid out on the floor in advance by the General Contractor, so we'll follow the layout even if it doesn't jibe with the prints. There are often changes made that aren't explained to us. And if the General Contractor did make a mistake, that's time and material — and T&M work is all good. We'll plumb up the plate line using a plumb bob or pocket laser and snap out the frame line. Now we're ready to plate the deck.

Step 2: Plate the Deck and Back Wall The deck will be plated following the standard top plate method, splicing the plate and staggering the pins from the front to the back edges approximately 24 inches apart. Shoot the plate up with either a gas- or powder-actuated nail set. The powder-actuated nail set shooting $1^{1/4}$-inch pins with yellow loads is the most common. Here we're screwing the top plate to a fire lid. We'll use drywall screws, hitting the ceiling framing when we can and using the cross tying method when we can't.

Your partner will help you get the first stick of plate tacked up. After that, it's a one-man job if you use the splices to hold up one end. If you also have a walk-up in addition to your scaffold, your partner can plate the back wall as you're working the top plate. Keep in mind, though, that your partner's main job is to keep you supplied with material.

As the back wall is plated, splice the plate together at the joints. Use drywall screws to screw the plate to the wall studs, two screws in each stud. After the deck's plated, plumb the plate line down the end walls and measure the stud length between the top plate and the bottom frame line elevation.

Step 3: The Studs In Chapter 4, we covered several methods you can use to calculate the stud length under both good and bad conditions. As we measured the stud length at the end walls of this soffit, we found the deck to be fairly level, so we subtracted 3/8 inch from the shortest end and cut all the studs the same length on the chop saw. Most foremen will give you a specific material to use when they line you out on a soffit. If not, use a stud length that'll give you the most soffit studs with the least amount of waste. The stud holes won't have to line up on a soffit unless the prints call for CRC to run through the studs. That's not the case here.

From the end wall, pull layout across the top plate as far as you can reach without moving the scaffold. By the time you're done, your partner will have at least a few studs cut to length and stacked on the scaffold. Begin with the end stud, stuffing it up tight in the top plate against the wall, and screw off the stud at both the front and back of the plate. Now set the stud to the plumb line and screw it off to a wall stud, or tie it into the drywall. From here, it's easy. Work your way across the soffit, stuffing the studs tight in the top plate on layout, screwing them off at both the front and back of the plate, as shown in Figure 11–15. As you work your way down the wall, pull the remaining layout from the soft side of the studs screwed off in the top plate.

Step 4: Kick It Off Many of the projects we've covered here used kickers to hold the work straight, with the kickers spaced 4 feet apart. Because the bottom of this soffit will be studded up, we'll use the bottom soffit studs to hold the soffit straight instead. This technique is also useful when you use DWC to frame in the bottom of the soffit.

Keep in mind that using the bottom studs to straighten the soffit (which we'll look at later) simply gives you another option. It's equally acceptable to pull a dryline and kick off a stud every 4 feet. Using the bottom studs is just another way to do the same thing. I assure you that you'll run into situations where one method is more suitable than the other.

Step 5: Plate the Studs Starting at the same end you pulled layout from, slide the end of the first stick of plate onto the end stud and clamp it in place. Let the plate run at a slight angle under the studs and work your way down the plate, twisting the studs and sliding the plate up onto them (Figure 11–16). Push the plate tight up on the studs and clamp it off every 6 feet or so. Once you slide the plate up on the studs, tack it to the studs that are clamped, then remove the clamps. Now pull layout across the bottom plate and set the studs to the layout marks. Next back out the tack screws, adjust those studs to layout and retack them.

Splice the next stick of plate to the first stick, and clamp the joint together. Using the same twisting method, slide the plate up onto the studs, clamping it in place as you go. When the plate is set, tack both the studs and the joint. Pull the stud layout for the rest of the soffit from

Figure 11–15. The soffit studs shoved tight up into the top plate and screwed off at both the front and back of the plate.

the soft side of a stud tacked on layout in the first stick of bottom plate.

Following this procedure, work your way down the soffit to the last stick of plate. Cut it to length and slide it onto the studs like the rest of the plate. At the end stud, set the plate to the frame line elevation and screw it off at both the front and back of the studs.

Step 6: Set the Bottom Plate To set the elevation of the bottom plate, snap a 3-inch chalk line from end to end of the soffit. Set the plate at the end stud to the bottom frame line elevation. Measure up 3 inches from the bottom of the plate and mark the end stud. Now run a tek screw into the 3-inch mark, leaving the screw hanging out far enough so you can hook the end of your chalk line onto it. When it's hooked, have your partner roll you down to the opposite end of the soffit, reeling out your chalk line as you go. Set this end of the plate to the frame line elevation and make your 3-inch mark on the end stud.

Snap the chalk line and you're ready to go. But take care. Pull the chalk line tight to avoid snapping a double line. If you pull the line off the studs at an angle when snapping, it will also result in a double line. Then you'll have to wipe off all the studs and resnap the line. Once the chalk line is snapped, have your partner wind up the chalk box and let it hang from the screw at the other end.

Work your way back across the soffit, setting the plate to the chalk line. Clamp three or four studs to the plate. Set the plate at each stud so that it's 3 inches below the chalk line, tapping it up or down with your axe. Then screw off all the studs you've got set. This process really speeds up when you work as a team, one partner setting the plate and the other screwing it off. As you come to the studs you tacked earlier, simply clamp the stud, back the screw out, and set it like the rest.

Figure 11–16. The bottom of the studs is being plated by twisting the studs as the plate is raised up onto them.

With the front of the plate set, the next task is to level the plate. There are two common ways to do this job. The first is to level the plate at each stud with a torpedo level, always at the bottom of the plate. The legs of the plate are often different lengths from side to side. The other method is to level each end of the plate and snap out another 3-inch chalk line, like we did to set the front of the plate. For this soffit, we've got room to work down the inside of the soffit, so we'll set the inside edge of the plate to another 3-inch line. The method will be the same.

Step 7: The Inside Plate Our next step is to plate the inside of the soffit to carry the bottom studs. Cut 2 feet off the first stick of plate to stagger the joints between the inside plate and the bottom plate of the soffit. Set the plate in place to the soffit studs and line up the bottom edges of the two plates. Then clamp the plate to the soffit studs, as shown in Figure 11–17. Screw off the inside plate with two tek screws at each stud, one at the top edge to the stud and one at the bottom, right beside the stud in the plate. Like all plate work, you splice the plate together, freeing up your partner to begin cutting the bottom studs.

Figure the stud length for the bottom studs by measuring the inside width of the soffit at each end, and allowing $3/8$ inch.

Step 8: Stuff Up the Bottom Plate Begin this step by pulling layout off the end wall down the front and back plate as far as you can reach without moving your scaffold. Your partner will have most or all of the bottom studs cut and stacked on your scaffold. You'll begin with the end stud, slipping it in the plate and sliding the hard side of the stud tight to the wall. In most cases, you'll be able to screw the end stud to at least one wall stud. If not, tie the stud into the drywall and screw the stud off at the top and bottom of both plates.

Work your way down the soffit from the end wall, stuffing the bottom studs on layout with the soft side of the studs facing the direction that layout was pulled from. Shove the studs tight into the back wall plate and screw them off at the top and bottom. Set the front of the studs on layout, and tack a stud every 6 feet. Screw the rest off when you set the front edge of the soffit.

Continue the stud layout, pulling off the soft side of the last stud you set on layout. Work your way down the soffit, stuffing all the bottom studs and tying in the opposite end stud. As you get to the end, you'll find it's often easier to work the end stud if you leave out the last layout stud until you get it tied in.

Step 9: Set the Outside Corner With all the bottom studs set on layout and screwed off to the back wall, we'll complete the soffit by setting the outside corner to a chalk line. This will plumb the soffit uniformly, just like we set the elevation earlier. At each end stud, measure back 6 inches and snap a line to the bottom studs, using a tek screw to hold the end of your chalk line. (You have to go back 6 inches to keep the chalk line on the studs.) Set the outside corner of the plate to 6 inches off the chalk line at each stud. Work your way down the soffit, clamping and setting several studs at a time, then screwing them all off.

As you work past the joints in the bottom (outside) plate which have been left unscrewed, set the plate to the 6-inch line on both sides of the joint. Then screw off the joint. Get all

Figure 11–17. The bottom inside plate clamped in place flush with the bottom plate. Once the plate is set, screw it off to the soffit studs.

Figure 11–18. You can see the bottom of the soffit stuffed up and screwed off, completing this drop.

the studs set and screwed off and this soffit's framed. See Figure 11–18.

Framing a Drop with Jigs

In this section, we'll put to work some of the techniques we covered in our jig discussion earlier, in *Tricks of the Trade*. We'll build the jigs for this drop out of $3^{5/8}$-inch material to form a light trough around the inside of the drop. We'll frame the drop to a level fire ceiling, and later a grid ceiling will be installed around the outside of the completed drop. Any time you're building with jigs, it's very important that you frame them to a relatively level surface for the jigs to be set at a consistent elevation. The grid ceiling elevations are also important. They'll determine which side of the drop you kick off the jigs from, and how far up from the bottom of the jigs you fasten the kickers.

As with any project, your first move is to clear the work area and stock your materials and equipment. In this situation, you'll also set up a jig table and a cut table. With our work area set up, we'll begin this drop by laying it out on the floor.

Step 1: Layout Use the gospel lines snapped out throughout the job to establish the top plate line. The top plate line will also be the leg of the jigs that supports them. It's very important when laying out the drop to make sure it jibes with any surrounding framework. In fact, laying out the drop is probably the most crucial step of the framing process. As you snap out the plate line, either snap both sides of the plate or mark an "X" on the side of the line the plate will sit on. Once it's snapped out, check the layout by measuring diagonally from corner to corner to make sure it's square.

Step 2: Top Plate The DWC of the suspended drywall ceiling framing we're framing this drop to runs only one direction. So two sides of the drop will run perpendicular and the other two sides will run parallel to it. The two sides of the drop that run parallel with the ceiling framing will require additional DWC and suspension wires to carry the drop.

Using your plumb bob or pocket laser, plumb up the layout lines to the fire lid close to the corner points of the plate line. As you're plumbing up the plate line, you'll also mark the 90-degree plumb points that you'll use later to establish the outside corner points. Again, mark which side of the plumb point the plate sits on with an "X" and the differences between the 90-degree plumb point and the outside corner point.

After you plumb up the first end of each side of the drop, run a screw into the ceiling at the plate line plumb mark. Hook the end of your chalk box to the screw and reel out your chalk line as you roll to the other end. Plumb up the other end and snap out the plate line. Then you can often pop the end of the chalk line off the screw by whipping the slack in the chalk line. Or if you can wrap the chalk box around the ceiling framing, or let it swing by the end hooked to the screw, use another chalk box to snap the next plate line. It's common to see "ceiling mechanics" with three or four chalk boxes in their tool buckets just for this purpose. You can save a lot of time by cutting down the number of times you have to roll back and forth, even when you're on a lift. Simply retrieve the extra chalk boxes as you're working the plate.

The first stick of plate will form the first outside corner and you'll use the 90-degree plumb mark to set it. When you plumbed up the plate line, you marked the 90-degree plumb mark and the difference between it and the outside corner point to the ceiling. Measure back from the end of the first stick of plate and mark this difference on the leg of the plate. Now just set the stick of plate in place. Line up the edge on the chalk line covering the "X" and the mark made on the plate on the 90-degree plumb mark.

Screw the plate to the ceiling, with two screws in each stick of DWC. The screws must run through the plate right at the outside edge into the ceiling framing when possible, without stripping out. Splice the plate together at the joints and overlap the corners, setting the first of the two sides of corners with the 90-degree plumb mark and cutting the second to overlap it.

Step 3: Building the Jigs In Chapter 4 we covered the procedures to build the jig itself. Following these suggestions, we built this jig out of $3^{5/8}$-inch material. With the jig built, determine the length of studs you'll need, and how they'll be cut. There's always an order you need to follow when putting the pieces of stud into the jig.

With the first of the jigs built, pull it out of the jig and check all the dimensions and the corners for square. Then make any needed adjustments to the jig. By the time you've got the jig built, your partner will be finished with the top plate. As one partner cuts the pieces of stud, the other puts jigs together and screws them off. Keep in mind this is a small-scale project we're building here. On large projects, two carpenters or even laborers will be assigned to just building the jigs after the jig is built.

Step 4: Set the Corner Jigs When the jigs face inside the drop to form a coffered ceiling, the corner jigs won't sit in the corner of the plate; their placement is determined by the overall width of the jigs. This will set up the inside corner of the jigs. A stud will be added later to form the outside corner. If, for example, the jigs are 14 3/4 inches wide, measure from the outside corner of the top plate 14 3/4 inches both ways and mark it. The corner jigs will sit on the far side of the marks. See Figure 11–19.

On the drop shown in Figure 11–20, the jigs face outside the drop so the corner jigs simply sit on each side of the inside corner of the top plate. With the first corner jig laid out, set the jig in the top plate and clamp it off. On this first jig, set the elevation by simply measuring down from the bottom of the jig to the floor. Once the elevation is set, screw off the jig in the top plate at both sides of the plate, making sure the jig is square in the plate.

Next, we'll plumb the jig down from the top plate and kick it off. On this particular drop, a grid ceiling will be installed around the outside. We'll kick off the jigs from the outside with the kickers running from the ceiling framing down to the jigs about 2 inches above ceiling height. Cut the kickers with one shoe that'll be screwed to the ceiling framing. Then screw the edge of the kickers to the hard side of the jig studs at the other end, with the kickers running at 45 degrees. Screw the shoe of the kicker to the ceiling framing and clamp the other end to the corner jig. Then plumb the jig with your level and screw it off to the kicker, as shown in Figure 11–21.

Since we're going to pull a dryline between the corner jigs, we'll run a second kicker down to the jig from the opposite direction (parallel with the top plate) to offset the tension of the dryline. Screw off the kicker to the ceiling and clamp it to the jig. Plumb the jig from the other direction and screw the kicker off to the jig. After the jig is plumbed and kicked off, set up the dryline at this jig and reel it out as you roll down to the other corner. Before you move, measure from the bottom of the jig to the center of the laser beam. Use this number to set the elevation of all the rest of the jigs. Move to the opposite corner, set the corner jig and pull the dryline.

Step 5: Kicker Jigs The jigs forming the drop must be kicked off every 4 feet. We'll work our way from corner to corner setting the kicker jigs, then filling in the layout jigs. Set the next jig on layout approximately 4 feet from the corner jig and clamp it in the top plate. Now set the kicker to the jig, clamp it off to keep it from resting on the dryline, and screw the kicker to the ceiling framing. Next, set the elevation with the laser, measuring from the bottom of the jig to the center of the laser beam. Screw off the jig in the top plate. Your next move is to set the jig to the dryline. Once it's set, double-check the elevation. When you've got it where you want it, screw off the kicker.

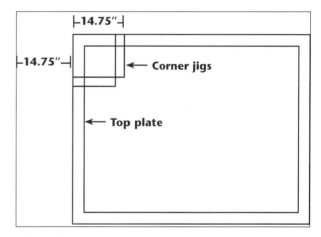

Figure 11–19. Setting the first two corner jigs 14.75 inches back from the actual outside corner point sets the inside corner of the jigs. The outside corner will be filled in later with a stud.

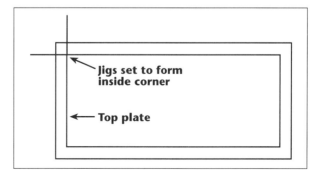

Figure 11–20. A drop framed with jigs facing out.

Figure 11-21. The edge of the kicker set to the hard side of the jig. Use your clamp-like pliers until the jig is plumb, then lock the clamp down and screw the kicker to the jig.

You'll use this process to set each of the kicker jigs as you work your way around the drop. Then set the layout jigs in between, held in place by the utility angle (which we'll set in the next step). As you work your way around the drop, you'll set the kicker jigs, utility angle and layout jigs in that order.

Step 6: Utility Angle The utility angle will tie the jigs together at the bottom corners of the jigs, both inside and out. In this step we'll set the utility angle at the bottom outside corner, between the jigs that are kicked off, then fill in the layout jigs. Load a bundle of utility angle and several kickers on your scaffold so you and your partner can gang up on the jigs.

Begin by setting the first stick of utility angle to the bottom outside corners of the corner and kicker jigs. Run the end of the utility angle wild about 16 inches past the actual corner point of the drop, and clamp it in place to the jigs. It's very important for the utility angle to fit squarely on the corner of the jigs. You can feel it when the angle locks onto the corner. When you've set the first stick of angle, screw it off to both the corner and kicker jigs at both the bottom and back of the jigs. Next, use your level to plumb the layout down from the top plate to the utility angle, and pull layout across the stick of utility angle.

As you work your way down the drop, overlap the joints in the utility angle by 6 inches. Clamp the joint right in the angle and screw off both ends of the joint on both sides of the angle. At the far corner, again run the utility angle wild out past the actual corner point. Once the utility angle is set and screwed off, use your level to plumb the corner point down from the top plate to the angle. After you turn the corner and get the corner jig set, cut the bottom leg of the angle right at the plumb mark, fold it around the corner and screw it off to the corner jig. Once the angle is tacked up from end to end, pull a dry-line right at the corner of the utility angle. We covered that in Chapter 4.

Your next move will be to set the layout jigs. Set the jigs in the top plate, letting them rest in the utility angle, and clamp the jigs off on layout. Now screw off the utility angle to the jigs on both edges, and set the elevation of the jig to the laser beam. Then screw off the jig to the top plate. See Figure 11-22. The jigs will follow the utility angle and automatically be set plumb.

In some cases, the utility angle will fall short of the corner. That's no big deal. Just run the utility angle on the opposite side of the corner wild and use the same method. That forms the corner you see in Figure 11-23.

Following these steps, work your way around the drop, setting all the jigs, as you can see in Figure 11-24. Once you're finished, it's time to move to the inside of the drop and get it knocked out.

Step 7: Plate the Light Trough Beginning at one of the corners, set the first stick of plate onto the ends of the studs to form the light trough. Here, because the studs are so short and tied into the jig, it'll be difficult to twist the studs into the

Figure 11–22. The first few jigs are in place with a jig kicked off every 4 feet. The utility angle screwed to the bottom corner will hold the rest of the jigs plumb as they're set.

Figure 11–23. If you look closely you can see the utility angle of the first side of this corner run wild. The second is side-folded and screwed off to it at the actual corner point.

Figure 11–24. The jigs are set in the top plate, kicked off and held in line with the utility angle.

Figure 11–25. The plate has been slid onto the corner and first layout jigs, set to width measuring off the utility angle and screwed off. The ends of both materials run wild — the plate to within 3/8 inch of the actual corner point.

plate. In situations like this, let the back leg of the plate slide onto the studs, then at each stud, pull the front leg of the plate out and down over the studs.

Now that you know how to get the plate onto the studs, eyeball the end of the first stick of plate 3/8 inch short of the actual corner point, and slide the plate onto the studs. Next, use your framing square to square the first layout jig from the utility angle to the plate, and clamp it off. From the layout stud, pull layout across the plate.

Before you move on, clamp the plate at the corner stud and set it to the proper width, measuring from the utility angle. Then screw it off, as shown in Figure 11–25. Like all light gauge plate, splice the plate together at the joints. Since we set the back of the jigs straight to a dryline earlier, this plate line will be set by measuring off the utility angle and we'll screw off the joints completely as we go.

As you reach the opposite end of this side of the drop, cut the last stick of plate 3/8 inch short of the actual corner point to form the outside of the light trough. You can also see this in Figure 11-25. Now roll back down this side of the drop, first clamping three or four studs on layout, then setting the plate to the proper width, measuring off of the utility angle and screwing the plate to the studs.

Once the front of the plate is set, roll back across the drop, leveling the plate at each stud with your level. Then screw off the plate inside the light trough. See Figure 11-26. That completes this side of the drop. As you turn the corner and start down the next side of the drop, turn the end of the first stick of plate into the end of the plate already set, forming the corner. Screw off the corner joints at each outside point. Following these procedures, work your way around the drop, setting all the plate. Figure 11-27 shows the completed framing.

(CRC) and suspension wires. Naturally, this will require the stud holes to line up.

For this example we'll use $3^{5}/_{8}$-inch 25-gauge material framed up to a corrugated metal deck. We'll be using a laser for all of our elevation work, but before we mount the laser to the wall, we'll use it to plumb up the wall line of the suspension wall.

In most situations, you'll do this work off of a lift. When using a lift, make sure first and foremost that you know what you're doing. If not, take the time to get proper

Figure 11-26. Using a level to plumb the face of the plate that forms the front of this light trough. Keep in mind that this very simple jigged soffit is part of a large drop.

Suspension Walls

Suspension walls are like a very large version of the vertical face of a soffit, and they're framed much the same way. The suspension walls break up large areas while still leaving them open. Malls use lots of suspension walls. If you've ever taken a good look at your local mall, you've probably seen an elaborate ceiling running down through the courtyard with a suspension wall on each side, separating the mall from the individual stores. The suspension walls will usually carry a lot of weight, which will be supported by cold-rolled channel

Figure 11-27. In the upper course of this multisection drop, you can see the completed light trough.

instruction. Secondly, keep the lift loaded with material so you can get up to the work and stay up. Going up and down for material unnecessarily is a big waste of time.

Step 1: Top Plate To begin the suspension wall, first establish and snap out the plate line on the floor. The proper layout of the suspension wall is crucial. The suspension wall will not only set the perimeter of the ceiling work, it'll also often set the lease lines for the tenants.

With the wall line snapped out, the next move is to plumb up the wall line to the deck. On this wall, instead of using a plumb bob, we're going to use the laser to plumb the wall line. That's common on long walls. First, set the laser perpendicular to the wall line and adjust it so the beam is centered on the line at one end of the wall. Your partner will work the opposite end of the wall to direct you as you dial in the beam to split on the wall line at the opposite end. Using a laser card will enhance the laser beam's visibility. As you dial in the laser beam, set the head of the laser in neutral and spin it by hand, speeding up a process that usually takes 10 to 15 minutes. This is time well spent though, especially when there are obstacles in the wall line. Once set, let the laser spin and roll down the wall marking plumb every 50 feet (or in between the bar-joist), then snap out the plate line.

As soon as you get the wall line plumbed up from end to end, put the laser back in its case and in the gang box. Next plumb up the corner points of the wall with your plumb bob, which completes the layout work for this wall. Now you can get it plated.

Work the plate following the standard top plate procedures. Since we're framing to a corrugated metal deck, screw the plate to the deck with S-12s, close to the outside edges of the plate.

Set the plate to the wall line and splice it together at the ends. Work your way down the wall, setting all the plate, and you're ready to stuff the studs.

The corrugated metal decking itself may present some problems. You can refer back to Chapter 1, *Shooting Up the Plate*. We covered there most of the common problems you'll run into.

Step 2: Stuff It Up For this step, load up your lift with studs which you'll run wild and cut to length later. You'll also load up some hanger wire. As one partner screws off the studs in the top plate, the other will tie the suspension wires to the top of the bar-joist.

Begin by establishing and pulling layout across the top plate as far as you can reach without moving the lift. The 25-gauge studs we're using have two stage holes punched in them. In all suspended walls like this one, the smaller of the two stage holes will always be on top. The CRC will later be run on edge through the small holes. The smaller holes prevent the CRC from sliding around. Clamp four or five studs on layout tight up in the top plate with the soft side of the studs turned to face the direction layout was pulled from. Then screw them off at both the front and back of the plate, as shown in Figure 11–28. Your partner will spend only a few minutes tying up the suspension wires, spacing them 4 feet apart, and then help you get the studs stuffed up and screwed off. Work your way down the wall setting and screwing off the studs on layout and tying up the suspension wires.

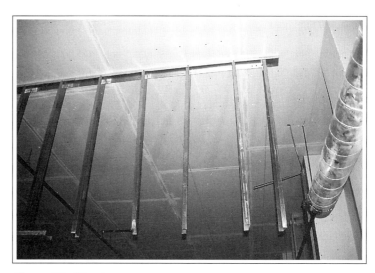

Figure 11–28. The suspension wall studs stuffed up tightly in the top plate and screwed off on layout.

Rolling down the wall on a lift may take a little getting used to. But on a flat level surface, like a concrete slab, it's common to move the lifts with the platform in the air. When driving the lift with the platform raised, keep your head out of the clouds and a close eye on your surroundings. Your partner will help watch for obstacles as well as watching your extension cords, which are easy to rip in half if you run over them.

Step 3: CRC and Kickers With the wall stuffed up, the next move is to lace up the studs with CRC, then get the kickers fastened in place at the deck. Lace the CRC through the row of stud holes closest to, but above, ceiling height. Just let the CRC lay in the stud holes for now. Our chief concern here is to get the wall laced up and the joints in the CRC screwed off with S-12s.

Cut the kickers long enough to run at a 45-degree angle down to about 6 inches above ceiling height. Here you'll only need to cut one shoe on the kickers. Screw that shoe to the corrugated metal decking with S-12s. Screw off the other end of the kickers to the hard side of the studs later, as they're set to a dryline. For now, just screw the kickers to the deck and tack them to the studs, or let them hang. Once the kickers are screwed off to the deck, you'll have the high work out of the way.

Step 4: Saddle Tie the Hanger Wires
With the high work completed, drop to the CRC elevation and work your way back down the wall, stretching and saddle tying the wires to the CRC. The method here will follow the same technique outlined in Chapter 3, except you won't have to worry about making the first bend at any specific elevation. In this situation, simply stretch the wire and cut it off about 16 inches below the CRC. Then pick the CRC tight up in the stud hole and tie it, as shown in Figure 11–29. As you tie the wires, eyeball the studs as close to plumb as you can. Then slide the wire to one side on the CRC and lock it down to remove any slack. Once you pull the CRC tight in the stud holes, it'll eliminate the need for a horizontal control stud.

Step 5: Set the Studs to a Dryline Beginning at an end stud, plumb and kick off the stud from both directions. Next, set up the dryline approximately 4 inches below the kickers, then roll down to the opposite end. Plumb and kick off that end stud and finish setting up the dryline. Follow the dryline method outlined in Chapter 4. Once the dryline is set up, work your way back down the wall, setting the studs every 4 feet to the dryline and kicking them off. See Figure 11–30.

Step 6: Cut the Studs to Length With the laser set up and running, establish the suspension wall plate line elevation. Next, figure the difference between the laser beam and the plate line. Then add $^3/_8$ inch for play and that's your cutoff point. Measure and mark the cut point on the studs as your partner follows behind you, cut-

Figure 11–29. The CRC held tightly in the smaller of the two-tier stud holes and the suspension wire saddle tied. Any slack is easily taken out by sliding the suspension wire to either side on the CRC and locking it in place.

Figure 11–30. The suspension wall studs being set to a dryline and kicked off.

ting the studs. The easiest way to cut the studs is to use a speed square to square across the hard side of the stud at the cutoff point. Then score the stud with your utility knife. Now cut the legs of the stud with your snips and fold the scrap end of the stud back and forth a few times until it snaps off. Once you mark all the studs you can reach, help your partner get them cut, as shown in Figure 11–31, then roll on.

Step 7: Plate the Studs Working from the same end you pulled the layout from at the top plate, slide the first stick of plate tight up onto the bottom of the studs and clamp it in place. On this suspension wall, we established the corner point of the wall and kicked the end stud from both directions when we set the dryline. That gave us an obvious starting and ending point. On many suspension walls, there isn't a beginning or end. The suspension wall is continuous around the entire project, so you'll simply pick a corner and go. Other times, the suspension wall will run from column to column and the end studs will be plumbed and tied into them.

With this corner established and the first stick of plate shoved tight up on the studs, tack it to a stud at each end and the middle. Once it's tacked in place, pull layout from the corner stud down the plate. Now splice the next stick of plate to the end and slip it up onto the studs, then tack it to the studs and at the joint. Hook the end of your tape to the soft side of a layout stud tacked in the last stick of plate and continue the layout. Following this process, work your way down the wall setting the bottom plate, as shown in Figure 11–32. Keep in mind that we'll be setting the bottom plate line elevation to the laser. Otherwise, we'd have set the elevation at the end and hook up a chalk line to reel out as we worked down the wall.

Step 8: Set the Plate Line Elevation Since we've got a laser set up on this project, we'll use it to set the bottom plate, saving a lot of time. Work your way down the wall clamping four or five studs on layout then setting the bottom plate, measuring from the plate down to the center of the laser beam (Figure 11–33). Once you've got each of the clamped studs set to the beam, screw it off to the front of the plate. As you and your partner develop a system, you'll learn to leapfrog over each other. One partner will clamp the studs on

Figure 11–31. The suspension wall studs being scored, then folded back and forth until they break off, forming a square cut.

Figure 11–32. With the studs cut 3/8-inch short of the plate line elevation, the plate is being slipped onto the studs and first clamped, then tacked, to the studs. With a laser setup, set the plate to the desired elevation as you go.

Figure 11–33. Here you see Mike measuring down from the bottom plate to the laser beam, setting the plate line elevation.

layout, then screw them off after the other partner sets the plate to the beam.

With the front of the plate set, clamp the back of the studs squarely in the plate and set the back of the plate, either to the laser or using your torpedo level. Work your way to the end of the wall, setting all the plate to completion.

F.Y.I.

Keep in mind that there are many variations in the conditions, techniques and even the order of the steps used to frame a suspension wall. Keep an open mind because there are always going to be hurdles to overcome. We covered many of them in the *Tricks of the Trade* section. The deck won't always be flat or even, or there will be obstacles in the wall line. The suspension wall may also come in many different shapes and sizes. Radiuses and multiple offsets are common. Figuring out how to deal with them is all in a day's work.

Another typical condition is a very small version of a suspension wall, or *bulkhead* as it's called. These walls are used to break up the intersections in grid ceilings. This is typical in corridor intersections. The method's the same — it's just on a much smaller scale. Here you'll simply use your level to set the bottom plate. Run the bulkhead right along the plate line of the corridor wall, preventing any offsets in the perimeter of the grid ceiling. In Figure 11–34, you can see the bulkhead completed and the grid ceiling installed. Notice the bulkhead extends only $5/8$ inch below the ceiling.

Figure 11–34. You're looking down a bulkhead that's framed to the fire lid with $3^{5/8}$-inch material. This bulkhead breaks up to grid ceilings to offset the different grid patterns of the two ceilings.

CHAPTER 12

Drywall Ceiling Systems

In this chapter we'll put our *Tricks of the Trade* (from Chapter 3) to work framing two common types of drywall ceilings, and a skylight. The first ceiling method we'll look at is a *suspended drywall ceiling,* where suspension wires support cold-rolled channel, which in turn carries the drywall channel that the drywall attaches to. The second, called a *hard lid,* is framed with studs. The *skylights* are also covered here because they're such a common part of commercial metal stud ceilings.

On any project, large or small, there's a sequence in which the work is done. When you're building the ceilings, the perimeter walls will usually already be framed, rocked and fire taped. This includes the suspension walls we covered in Chapter 11, and the walls of the skylights we'll cover later.

Like all the examples we've worked through, keep in mind that the ones in this chapter are very basic framing projects. Once you've mastered them, you can just expand on them as your projects get more complicated. A metal stud ceiling will go up very fast. That's good — unless that speed makes you miss openings in the framing for the other trades. So let me again impress upon

Commercial Metal Stud Framing

you the importance of *thoroughly studying* the prints *before* you start the work. Most of us, when we start out in the trade, want to just get on with the job. Who has the time to waste poring over plans? Take my advice and take the time. Remember, there's no profit in rework.

Suspended Drywall Ceilings

There's no limit to the size of a suspended ceiling. They range from a single room to the ceiling of an entire shopping center. An architect's imagination is the only limit as to how elaborate and intricate a ceiling may be. These complex ceilings, full of angles, offsets, domes and arches, can be real head-scratchers, even for the most seasoned ceiling mechanic. But for this exercise, we'll work through the process of building a straight, flat suspended drywall ceiling.

There are a couple of different material systems you can use. The first system covered here is the more common of the two, largely because it's so versatile. This system uses hanger wires to support the cold-rolled channel (CRC) which in turn carries the drywall channel (DWC) that the drywall is screwed to later. The second system goes together just like a grid ceiling, using *mains* that are bridged with cross tees or cross channels. The entire system simply snaps together. The Rigid X system is very fast, but the materials are more expensive, so you just don't see it as much as the first system we'll look at here.

As we work through these steps, keep in mind that the ceiling work is usually done off of lifts. It's also common for two partners to split up and knock out two steps simultaneously. So let's put this ceiling system to work.

Step 1: The Elevation While a water level will do the job on a small- or medium-size ceiling like this one, we'll use a laser because that's the most common way to frame a suspended ceiling. To begin, get the laser set up on one of the perimeter walls. First establish the ceiling frame line elevation off the floor, then the difference between it and the center of the laser beam. Now work your way around the perimeter walls, marking the elevation in the corners and snapping a chalk line corner to corner. Remember, there's a limit to how far you can accurately pull a chalk line. Walls over 50 feet long need additional elevation marks. When all the lines are snapped out, you're ready to install the utility angle on top of the chalk line.

Step 2: Utility Angle Starting at any corner, set the utility angle to the chalk line and screw it off to the wall studs with drywall screws. As you're doing so, make certain that the screws don't strip out. That's likely to happen when you screw through the utility angle and the drywall into the studs. On light gauge studs, you might have to work the drywall screws into the studs slowly. That's something you'll quickly get a feel for. In our example, the perimeter walls weren't rocked in advance, as they usually are. We'll just use framing screws instead of drywall screws to fasten the utility angle to the studs.

Overlap the joints in the utility angle by at least 6 inches. When the joints are close enough to a wall stud, overlap the joint far enough that you can screw the ends of both sticks of utility angle to one stud. Otherwise, you can simply screw off the joint with tek screws and tie it into the drywall just like you tie in a slap stud. Work your way around the perimeter walls, screwing all the utility angle to the walls. See Figure 12–1.

Step 3: Hang the Suspension Wires Using a wire tier, tie the suspension wires to the tops of the bar joist. We covered that procedure in Chapter 3. Hang the wires in neat, even rows 4 feet apart. Maintain a 4- by 4-foot pattern, beginning 4 feet off the wall. The 4- by 4-foot pattern lets you hang the CRC in straight 4-foot rows perpendicular to the bar joist.

Step 4: Stretch and Bend the Hanger Wires With the suspension wires hung, your next move is to stretch and bend every fourth wire in each row. The rest will be stretched and saddle tied after the CRC is up. To begin, establish the bend point. This will be $1\tfrac{1}{8}$ inch above the frame line elevation that you set the utility angle to. That $1\tfrac{1}{8}$ inch comes from the DWC, which is $\tfrac{7}{8}$ inch plus $\tfrac{1}{4}$ inch for play. The play is taken up as the ceiling is "dialed in" later.

The actual stretching and bending process was covered in detail in Chapter 3, so we'll skip the mechanics here. Work your way across the

Chapter 12: Drywall Ceiling Systems   173

Figure 12–1. Working around the perimeter walls, screwing off the utility angle to the wall studs. As you can see, these walls haven't been rocked yet. That's fine. We're simply using framing screws instead of drywall screws to screw the utility angle to the wall studs.

Figure 12–2. The rows of CRC tied up just under the bar joist are held straight with a single stick of DWC. A few suspension wires are still hanging straight and will be tied after a few sticks, or sometimes all the DWC, are tied up.

entire ceiling area, stretching and bending every fourth wire in each row, as shown in Figure 12–2. Bending every fourth wire will give each stick of CRC two bent wires to carry it as you hang the CRC, which is your next step.

Step 5: Hanging the Cold-Rolled Channel As you prepare to hang the CRC, the first move is to establish the 4-foot layout for the CRC along the end wall where you're starting. Then pull layout as far as you can reach down the wall without moving your scaffold. At the utility angle, the CRC must stay $7/8$ inch above the frame line elevation to allow for the DWC.

There are a couple of common ways you can do this. The first is to cut a 2- or 3-inch shoe on the end of the CRC and tie it into the wall $7/8$ inch above the utility angle (Figure 12–3). The second method, which we'll use here, is to cut some 1-inch pieces of DWC on the chop saw and screw them to the utility angle at the 4-foot layout marks. Then you screw off the end of the sticks of CRC (first and last of each row) on layout to the pieces of DWC, as shown in Figure 12–4.

Beginning with the first stick of CRC, bump it up against the end wall and set it in the bend of the suspension wires. With the CRC resting in the bend of the wires, set the end of the CRC on layout, and screw it off to the short piece of drywall channel with a single 34. Next, eyeball the wires plumb, and tie them with the saddle tie method (also covered in Chapter 3). After

Figure 12–3. Here you can see 2-inch shoes cut on the end of the sticks of CRC, which are then screwed off to the wall studs.

Figure 12–4. The CRC held up $7/8$ inch above the ceiling frame line with a small piece of DWC. This is the most common method of holding up the CRC.

Commercial Metal Stud Framing

you've tied the prebent wires to each stick of CRC, stretch the remaining wires and saddle tie them to the CRC. Keep in mind that some ceiling mechanics prefer to leave the unbent wires hanging, to stretch and tie them after the drywall channel has been fastened to the CRC. One method's as good as the other. It's a matter of personal preference — something you and your partner will have to agree on.

Set the next stick of CRC in the bend of the wires and overlap the CRC joint by approximately 12 inches. Then saddle tie the prebent wires. In Chapter 3 we discussed the two common ways to fasten the joints in the CRC. In this case, we'll fasten the joints with tie wire. At the opposite end of each row, cut the CRC to length and hang it like the rest. The opposite end wall will also be laid out 4 feet on center, using short pieces of DWC to maintain the elevation of the CRC at the wall. Following these procedures, work your way across the entire ceiling area, hanging all the CRC, as shown in Figure 12–2.

Step 6: The Drywall Channel The next step is to fasten the DWC (or *hat channel*) to the CRC. For our sample project we'll use tie wire to tie up the DWC. To get going, load up your lift with several bundles of DWC. Lay out the bundles of DWC 4 feet on center along the edges to keep the rows of CRC straight. This method is pretty simple: just square up one end of the bundles and pull a 4-foot layout along the edges, marking the edges of all the DWC in the bundle with your felt-tipped marker. Keep in mind that you'll have to reestablish the 4-foot layout after the first row of DWC to allow for the overlapping of the joints. With your first row of bundles of DWC laid out, it's time to get up in the air.

Once you're up there, pull layout down the utility angle, measuring off the end wall. For this ceiling, like most ceilings, the DWC layout is 16 inches on center. Once you've laid out the utility angle as far as you can reach, you're ready to lay out the CRC. If you marked the centers when you pulled layout across the CRC, the DWC would cover your marks — so pull layout for the edges. The DWC is $2^{1/2}$ inches wide edge-to-edge, so make your first layout mark at $14^{3/4}$ inches. You're subtracting half the width of the DWC ($1^{1/4}$ inch) from the 16-inch layout. From this mark, pull your 16-inch layout across the CRC.

To pull an accurate layout across the CRC, you'll need to "walk" your tape across the crooked (snaking) CRC. It's also common to tack up a stick of DWC every 8 feet and set the CRC to the 4-foot layout marks you made earlier on the DWC to straighten out the CRC. Then you can pull an accurate layout. You don't need to lay out every row of CRC. Each stick of DWC will hit at least two rows of CRC. Pull layout across the furthest row each time. It usually works out to every other row. Remember, as you're laying out the CRC, that you'll be marking the *edges* of the DWC. Put an "X" on the side of the layout marks that the DWC will sit on.

Once you've completed all the layout work that's within reach, you're ready to start setting and fastening the DWC in place. The DWC will be tied to the CRC on this ceiling following the tie wire procedure covered in Chapter 3. Screw the DWC to the utility angle using tek screws. When you have two partners working off one lift, one will tie and the other will screw off, then help tie as time permits. You'll quickly develop a system.

In Figure 12–5, one mechanic is hanging the DWC alone. In this case, set up the sticks of DWC in the utility

Figure 12–5. We're working the ceiling off of stilts. Steve is working his way across the ceiling, setting the ends of the DWC up into the utility angle on (or close to) layout, then fastening the DWC to the CRC on layout.

angle close to layout, with the end tight to the wall, clamping a stick every 4 feet, then tie it to the CRC on layout. Don't cinch the wire down, just tie it to the laid-out row of CRC. Then set the next stick of DWC and so on, until you've tacked up all the DWC you can reach, lining up the layout marks on both the materials as you go. Then work down the utility angle, screwing off all the sticks of DWC on layout. After the ends are screwed off, add the tie wires to the middle row of CRC again, setting it to the 4-foot layout marks on the DWC (Figure 12–6). Finally, start tapping the materials to their exact layout marks with your end nippers and cinching down the wires, as shown in Figure 12–7. Work your way down the ceiling to the opposite wall, tying up all the DWC in this first run.

To start the next run, first establish the 4-foot layout on the bundles of DWC. Measure back from the end of a stick of tied-up DWC to the CRC it's tied to. Let's say it's 24 inches. Now square up the ends of the bundles and make your first layout mark at 8 inches. This will mark the overlap for the rows of DWC. From the 8-inch mark, burn that 24 inches and mark the 4-foot layout to the edges of the bundle of DWC.

With the DWC laid out, tack a couple of sticks in place to straighten out the CRC so you can lay it out. The ends of the DWC in this run will sit up in the ends of the DWC in the last run, clamping the joints. Then tie the opposite end to the row of CRC you laid out, lining up the layout marks on both materials. Once you've got several sticks of DWC tacked up, go back through, tying off the joints and lining up the 8-inch mark on the ends of the first run of DWC. Tie the joints in the DWC at each end of the joint with a single strand of tie wire that you've doubled up. As you tie off the joints, add the wires to the middle row of CRC, just hand tying all the wires for now. After all the DWC within reach is tied up, go back through and cinch down all the wires. You can see that in Figure 12–7.

Following these procedures, work your way across the ceiling, tying up all the DWC. Cut the last stick of DWC in each run to length and screw it off to the utility angle on layout at the opposite end wall.

Step 7: Dial in the Ceiling Work your way across the ceiling, setting the rows of CRC to their proper elevation. Measure down from the DWC (closest to each suspension wire) to the center of the laser beam. While you're

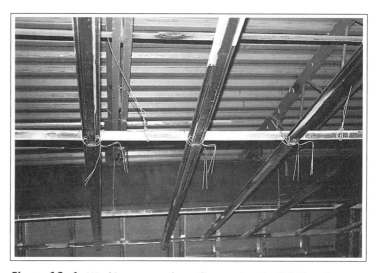

Figure 12–6. Working across the ceiling setting the DWC on layout and hand-tying the DWC loosely to the CRC.

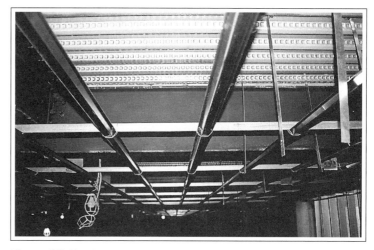

Figure 12–7. The tie wires are cinched down, completing the tying process. Notice the straight, even rows of both the DWC and CRC, a result of properly laying out both materials.

measuring, pull down on the CRC at the suspension wire to imitate the weight of the drywall that will be hung later. Adjust the elevation by sliding the suspension wire either way on the CRC until the DWC is at the same elevation as the utility angle, as shown in Figure 12–8.

As you get each wire adjusted, lock it in place. Do that by biting onto the bottom lip of the CRC with your linesman's pliers and bending a hump in the CRC that'll keep the wire in place. Figure 12–9 shows the result. You can also lock in the suspension wires by running an S-12 into the CRC right next to the wire. Work your way across the ceiling, dialing in all the suspension wires, and you're done.

Keep in mind that this example was about as simple as a suspended drywall ceiling can be. There were no light or vent openings, no angles or changes in elevation. But once you master the basic mechanics, you can build on them for the more complex ceilings.

The Rigid X System

Before we finish up our discussion on suspended drywall ceiling systems, I want to lightly touch on a second material system used to frame a ceiling. The Rigid X system uses a grid-ceiling-style material instead of the heavy gauge cold-rolled channel to support the weight of the drywall. That weight can be considerable, especially if the drywall's applied in multiple layers to achieve up to a 3-hour rating. This system uses a variation of the same materials used in normal suspended ceilings. But in the Rigid X, most of them are factory painted so the ceiling framework can be left exposed. Then the drywall is laid in on top of the ceiling as ceiling panels, just like a grid ceiling, or screwed to the framing. Let's look at the components you'll use in a Rigid X system.

If the drywall will be screwed directly to the bottom of the framing, you'll use the same utility angle used in our first system. But if the drywall is installed as lay-in panels, you'll use a prepainted wall mold instead of utility angle. No matter which angle you use, you'll screw it off to the wall studs just as in our first ceiling system.

Figure 12–8. Pulling down on the CRC, pulling the slack out of the suspension wires as the elevation's set to the laser.

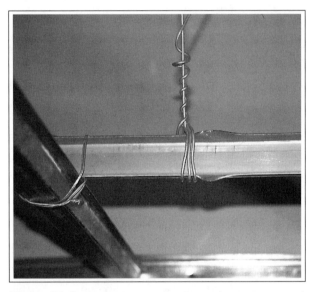

Figure 12–9. Once the elevation's set, the CRC is spread open right at the saddle tie, locking in the ceiling at the elevation.

Tie the suspension wires to the tops of the bar joist and space them 4 feet apart. The difference in procedure is in stretching and bending the wires. Once stretched, bend the wires while they're lined up with the prepunched holes in the mains, as you set the main to the laser. While this may sound like a circus act, it's really pretty simple. First set the main up in the wall mold and clamp it on layout. Let the wire slide through the fingers of your left hand (assuming you're right handed) while holding the main tight to the wire with your index finger. Line up the wire with the hole it will run through. Set the elevation, then eyeball the hole to the wire and bite under that point with your linesman's pliers and bend it sharply horizontal. See Figure 12–10. Then slide the wire through the hole and wind it tightly four or five times.

The mains (which replace the CRC in our last ceiling system) are 12 feet long with holes laid out and prepunched for both the suspension wires (4 inches on center) and the cross tees (and channels). The ends of the mains snap together, which automatically maintains the layout for everything all the way across the ceiling. Lay out the mains along the utility angle (or wall mold), maintaining a 4-foot on center layout running perpendicular to the bar joist (Figure 12–11). Unlike the cold-rolled channel in our previous ceiling, the mains will hang at the same elevation as the utility angle. The square ends of the mains are recessed to allow the mains to rest in the angle without any offset. Set the ends on layout, then screw them off to the angle when the drywall will be screwed to the bottom of the framing. If the rock is laid in, pop rivet the ends in place. When planning the Rigid X system as a lay-in ceiling, establish the layout for the mains from the center of the room out, so the lay-in panels are even on the two outside edges of the ceiling.

The mains are bridged with either cross tees or cross channel, both of which are 4 feet long and simply snap into the prepunched holes in the mains. The cross tees are made from the same grid-ceiling-style material as the mains. You can use them in a lay-in ceiling. When screwing the rock to the Rigid X framing, you'll use cross channels made of DWC-type material. Figure 12–12 shows the end of a Rigid X cross channel.

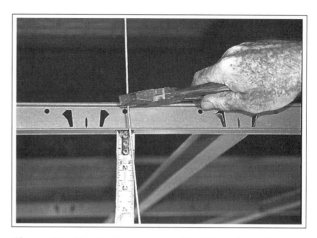

Figure 12–10. Pushing up on a Rigid X main as it's set to the laser, then grabbing the prestretched suspension wire right at the hold it'll be tied to.

Figure 12–11. The ends of the mains screwed off on layout to the utility angle. In this shot you can also see the cross channels snapped into the mains, maintaining a straight, square grid pattern.

Figure 12–12. The end of a Rigid X cross channel, clearly showing the hooks which snap into the mains.

Hard Lid Ceilings

A hard lid is another common method of framing a drywall ceiling. It uses metal studs (usually $3^{5}/_{8}$ inches) instead of suspension materials to frame in the ceiling. This method is used in small- to medium-sized rooms where the metal studs can span the room wall to wall. As with any project, before you can frame in the hard lid, the perimeter walls of the room must be rocked and fire taped above the ceiling.

For this example we'll frame a ceiling out of 3⅝-inch 25 gauge studs, 8 feet off the finished floor. The ceiling will later be rocked with one layer of ⅝-inch rock, giving us a frame line elevation of 8 feet ⅝ inch.

Step 1: Plate the Walls To begin, establish the ceiling's frame line elevation in one of the corners. Use benchmarks made with your water level or pocket laser to transfer the elevation mark around the room in each corner. Then snap out the frame line from corner to corner, making sure the chalk line is pulled tight.

The plating process is worked much like the utility angle of a suspended ceiling. Beginning in any corner, set the first stick of plate on the chalk line, pushing the end tight into the corner to hold it up. Screw off the plate to the wall studs with two drywall screws in each stud, as shown in Figure 12–13. The ends of the plate will be spliced together, so stay at least one stud back from the end when screwing the plate to the wall. Use the splice to hold up one end of the next stick of plate as you set it to the chalk line and screw it off. Then screw off the joint in the plate with a couple of tek screws.

As you reach the opposite corner, cut the last stick of plate ¼ inch short, then screw it off. After you turn the corner, slide the first stick of this wall to the last stick of the wall just completed to hold the end in place as it's screwed off. Continuing this process, work your way around the perimeter of the ceiling, screwing off all the plate.

Step 2: Install the Studs The studs will span the narrowest distance across the room, so begin stuffing up by pulling layout down the plate off the end wall. Lay out just one side of the room for now, laying out as much of the plate as you can reach without moving the scaffold. Cut the studs to length when necessary, then reach the end of the studs across the room and set the ends in the plate. See Figure 12–14. As the end of each stud is set up on the far plate, twist your end of the stud in the plate, then set the stud on layout and screw it off. As with all metal stud framing, stuff the studs with the soft side of the studs facing the direction layout was pulled

Figure 12–13. Setting the 3⅝-inch plate to the plate line and screwing it off to the wall studs.

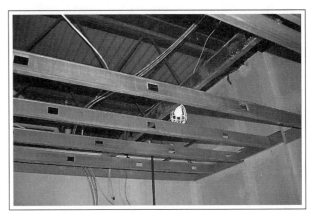

Figure 12–14. Once the perimeter walls are plated, the plate is laid out and the studs screwed off on layout as far as the carpenter can reach without moving.

from, then screw them off on both sides (top and bottom here) of the plate.

Step 3: Add the Stiffback Once you've stuffed up about half of the studs, cut your stiffback (which we covered in Chapter 2) to length and set it up on top of the ceiling studs. You could set up the stiffback in the ceiling after all the studs are stuffed up, but it'll save a lot of time and hassle if you set it up in the ceiling ahead of time. With the stiffback in place, finish stuffing up the studs and screwing them off on your side of the ceiling.

Then move across to the opposite plate and establish layout. Work your way down the plate, pulling layout and screwing off the studs. That finishes the stuffing up process.

Now set the stiffback on edge in the center of the ceiling and screw it off. Screw off each of the ceiling studs to the stiffback with tek screws, running the screws up through the top lip of the ceiling studs into the stiffback, as shown in Figure 12-15. To get the most support out of the stiffback, run the screws through the studs close to the hard side of both. Screw off the stiffback — and this ceiling's finished.

Penetrations in the Hard Lid

Before we wrap this up, let's take a quick look at the same hard lid with a rough opening framed in it. Lay out the rough opening along the plate as you pull the stud layout on each side of the room, marking the inside of the rough opening. The rough opening in the hard lid will be framed much like the rough opening in a metal stud wall. Set the first studs of the rough opening in the plate with the hard side of both studs set to the layout marks. Then screw them off. Next, establish the other two sides of the rough opening, marking the inside on the first two studs. Span the studs with two headers, again setting the hard side of the headers to the layout marks and screwing them off. Complete the rough opening by adding any needed cripples and screwing off the stiffback, which runs just off to the side of the rough opening.

Skylights

For centuries, skylights have been used to bring light indoors. With the introduction of metal studs, it became much more economical to transform the skylights from a large square hole in the ceiling to a work of art. Here we'll frame in a simple square skylight. Remember that we covered the details of each example earlier, in Chapter 4. The most important step is fastening the metal stud framing to the red iron. Once the skylight is framed and rocked, it'll weigh several hundred pounds, and must be completely self supporting.

Step 1: Layout Like all of the framework on the job, you'll establish the skylight layout off of the gospel lines, or the frame line of other work in the immediate vicinity. With skylights, however, space is usually the biggest issue. You'll only use the gospel lines to establish reference points at each end of the skylight to ensure it's square with the surrounding work.

Figure 12-15. With several of the hard lid studs screwed off on layout, we get a better view of the stiffback set up on top of the ceiling studs. The stiffback is centered up and screwed off to the top leg of the hard lid studs.

To figure out this skylight, first plumb up the gospel line from the floor and make a reference mark at each end of the skylight rough opening. Next, measure the overall rough opening of the red iron at each end and establish the center of each. Now check the difference between the reference and center marks at each. In an ideal situation, the difference would be the same at each end. But the red iron is rarely close to where it's supposed to be, let alone perfect. Here we've got $1/2$-inch difference from end to end. That tells us that the skylight rough opening is a little out of square. From here it's a judgment call. Set one end to the actual center point, then use the reference marks you plumbed up earlier to establish the center point at the opposite end.

Since the rough opening is out of square, we know one corner is going to be narrower than the rest. Establish the narrow corner, then measure out of the narrow corner $3^{3}/_{4}$ inches (since we're using $3^{5}/_{8}$-inch material) and mark the end wall. This is your frame line. Use the reference mark to transfer the point to the other three corners of the skylight. Using your 4-foot level, plumb the points up the red iron. Using the same method, establish the frame line for the two long sides of the skylight, making sure the frame line width is the same at each end. The end walls will be squared off the long walls once they're plated and stuffed up, using the 3-4-5 technique.

Step 2: Shoot the Braces to the Red Iron Remember, the completed skylight is going to be pretty heavy, so you've got to brace the framing solidly. To carry the framework, we'll use a simple stud brace (from Chapter 1) to carry a layout stud every 4 feet. The plate will carry the rest of the layout studs. Cut the braces approximately the same length as the red iron is wide. You don't want the brace to extend above or below the red iron.

As your partner cuts the braces to length, establish the layout for the skylight studs and plumb it up and down the red iron. Figure the hard side of the studs, and pull layout across the red iron. Mark the 4-foot studs and corner studs 6 inches out of the red iron corner on each long wall, at both the top and bottom of the red iron. Shoot the braces to the red iron on edge, with the hard side of the braces on the opposite side of the layout marks from the skylight studs. This will allow you to screw the studs to the braces hard side to hard side.

Shoot the braces to the red iron beginning at an end wall. Set the hard side of the brace to the red iron with the edge of the brace lined up on the plumbed frame line you established earlier. Shoot the braces in place with steel pins and red or yellow loads, as shown in Figure 12–16. When you're cutting the braces from light gauge material, it's easy to force the barrel of the shotgun into position to the inside leg of the brace. Using a hat channel head will make it even easier. When cutting the braces out of heavy gauge material, have your partner hammer one leg flat as they're cut.

Step 3: Set the Studs On this skylight, we're running the studs wild past the bottom frame line, so our only concern for the elevation of the studs is that they're set roughly $3/_{8}$ inch below the red iron in the top of the skylight. Later, we'll set the top plate flush with the red iron. The rockers will screw their rips to the top plate on one edge and laminate the other to the red iron.

With this in mind, the first studs we'll set are the corner studs of either of the long walls. First complete one wall, then frame the other. Clamp the studs to the corner braces one at a time, eyeballing the tops $3/_{8}$ inch below the red iron. Next, set the edge of the stud to the frame line you established earlier and plumb the stud with your 4- or 6-foot level. Once it's set, screw the stud to the brace with two screws in each brace. Before rolling down to the opposite corner, set up a dryline at both 3 inches and 4 feet (approximately) down from the top of the corner stud. Move to the opposite end and set the corner stud, then pull the drylines, but not so tight that they pull the corner studs in out of position. Another method is to pull only one dryline and use a level to plumb the studs as they're set. Use whichever works best for you. I use two drylines because if the studs are just a little out of plumb, the flaw would be cantilevered to the bottom plate, causing dips or humps in the plate line. That's the most noticeable line of the skylight. On shorter skylights, run your top plate between the corner studs to keep the tops straight, and run a single dryline at the bottom of the studs.

Figure 12–16. A clip cut from $3^{5}/_{8}$-inch stud shot to the red iron superstructure with two steel pins. A skylight stud is plumbed and screwed off to the clip with two 34s.

Work your way across the wall, clamping the studs to the braces with the tops approximately 3/8 inch below the red iron, then setting them to both drylines. As the studs are set, screw them off, working your way down the wall, as shown in Figure 12-17. Complete the first long wall, then plate the top of the studs. That's our next topic. Once the first long wall is completed, work the second long wall.

Step 4: Top Plate On this skylight, the top plate runs along the inside of the red iron angle. See Figure 12-18. In this situation you can simply cut the top plate to length and slip it onto the studs as they're set to the red iron. On skylights where the top plate forms the top of the skylight, set the plate to a 3-inch line, maintaining a straight, level top plate.

For some skylight walls, you'll complete the stuffing up phase by adding a horizontal control stud to support the weight of the unbraced layout studs. The need for a horizontal control stud depends on the length and subsequent weight of the unbraced studs. If left unbraced, would they pull dips in the top plate? I use one any time the studs are 10 foot or longer. Run the horizontal control stud across the skylight studs, screwing it off first to the braced studs, then to the layout studs as they're stuffed. One tek screw per stud will do the job. Be sure the unbraced studs maintain layout. As the walls of the skylight are rocked, the horizontal control studs will be removed. Follow these procedures to get the two long walls stuffed up and we're ready to move on to the end walls.

Step 5: Cut the Studs to Length Your next move is to cut the studs to length (or in this case, height). Cut them 3/8 inch above the bottom frame line elevation to allow for some play when you set the bottom plate. Establish and maintain the elevation of the cut point by measuring up from the laser beam. Work the skylight in sections, marking as many studs as you can reach without moving your lift. Your partner will come behind you scoring the hard side of the studs squarely, using a speed square and utility knife. Once the studs in the section are all scored, cut both legs of the studs and fold the scrap end back and forth a couple times until it snaps off. See Figure 12-19. Work your way around the bottom of the skylight studs and get

Figure 12-17. Here the skylight studs are being set right to the steel, simply following the red iron. This is common.

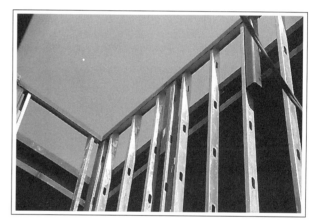

Figure 12-18. These skylight studs come up inside the red iron angle. They were cut to fit in between the red iron support posts, then added as the studs were set.

Figure 12-19. Using a laser, establish the plate line elevations, and cut the stud after allowing 1/4 to 3/8 inch for play. Here you see the hard side of the studs scored and the legs of the stud cut. After the legs are cut, the studs are folded back and forth until they snap off.

around the bottom of the skylight studs and get them all cut to length.

Step 6: Bottom Plate   With the studs cut to length, you're ready to get the bottom of the studs plated and the plate set. You want to begin at one of the corners but there isn't any outside corner stud in place yet. So straightedge the end of the plate to within 1/4 inch of the corner point. Twist the bottoms of the studs one at a time and slip the plate up on them. Splice the joints together in the plate and overlap the corners. Work your way around the skylight, tacking all the plate in place.

Once the bottoms of the studs are plated, set the plate line elevation to a 3-inch chalk line established off of the laser beam. Setting the studs to a dryline and kicking them off isn't necessary because you plumbed and braced the studs from up inside the skylight. Work your way around the bottom of the skylight studs, laying out and setting the bottom plate. Set the plate, clamping off four or five studs on layout at a time, and setting the front edge to the 3-inch chalk line. Then screw them all off. Set the inside of the plate with a torpedo level, screwing the plate to each stud as it's set. Figure 12–20 shows the completed project.

Figure 12–20.   The bottom plate is set, completing the skylight.

Chapter 13

Drywall Methods

As a metal stud framer, it's inevitable that you'll spend a little time hanging drywall. That's why I'll cover some of the basic tricks of the drywall trade. Hanging drywall isn't only physically demanding — there's considerably more to it than most people think.

Most of the rock on commercial jobs is applied vertically (stood up), which eliminates most of the butt joints common in horizontal installations (laid down). Keep in mind that the drywall work must be neat, with tight joints and even screw patterns, regardless of which way you apply the sheets. You can avoid other frequent mistakes by planning ahead. When you stand up the drywall, don't let the joints on either side of a wall break on the same studs. When you lay down the rock, stagger the joints by at least 32 inches. For long walls that are laid down, snap a reference line at $48^{1}/_{4}$ inches up off the floor, and lift the sheet to the line to keep the rows even.

Now let's cover some tricks of this trade—tricks that can make the difference between making decent money and just scraping by.

Marking Sheets with a Tape Measure and Pencil

This method will help you to eliminate a lot of wasted time spent chalking a line. A variation of this method, using your knife instead of your pencil, is another big time-saver. Each of these methods is pretty straightforward and simple, but takes a little practice to master. Once you have this trick under your belt, you'll work more efficiently — and probably make more money.

Begin by first measuring to and marking the point of the line. With practice you'll be able to just pull your tape out to the desired width, and match the measurement on your tape to the edge of the sheet. Let's draw this line at 16 inches down the length of the sheet. Measure down 16 inches from the recessed edge of the sheet and mark it. Now hold the point of your pencil to the end of your tape measure, with your thumb on the inside of the tape end, and your index finger on the outside of the pencil, pinching the two together, as in Figure 13-1. The side of your other index finger will act as a guide, sliding along the edge of the sheet, so make sure it's clean.

With your tape in hand, set your index finger to the edge of the sheet, and pull the end of your tape and pencil down to the mark. Now *slowly* slide both hands down the sheet evenly, as shown in Figure 13-2. It's important to keep your hands even as you slide down the sheet. If you don't, the line will waver as your hands come out of square with each other.

That's it! Now just practice a little and you'll have it mastered in no time. This method will work from either the recessed edge (seam) or the butt (end) of the sheet.

Cutting Drywall

Before we get going, let me remind you about something important — your knife. Treat it with respect. Let your attention wander when you're cutting and you'll learn respect in a hurry. A sharp blade is safer than a dull one. I've given myself lots of cuts in my time, and the cuts from a dull blade are almost always worse than the ones from a sharp one. Possibly because I'm pressing harder.

Cutting with a Utility Knife

We'll begin by using your utility knife with a tape measure. The method is very similar to using your tape and pencil. Again, we'll make a

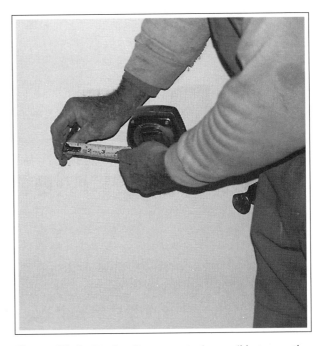

Figure 13–1. Pinch a flat carpenter's pencil between the index finger and thumb as it's held to the end of the tape.

Figure 13–2. Slide one hand with the tape case along the top of the sheet, drawing a straight line with the pencil pinched to the end of the tape. Keep the tape square with the edge of the sheet to maintain a straight line.

16-inch rip. To begin, mark your sheet at 16 inches. Hook the blade of your utility knife to the inside end of your tape measure (Figure 13–3). With the tape case in your other hand, use your thumb and index finger to pinch the tape and hold it where you want it. You're using your index finger as a guide as you make the cut. Set your finger to the top edge of the sheet and pull the blade of your knife and tape down to the mark. Now slide down the sheet from end to end, making the cut. In Figure 13–2, you can see a pencil line that marks where you'd make the cut. Remember to keep your hands square. This method also takes some practice to perfect. Find yourself a piece of scrap and cut it to pieces.

A T-square is another good way to make a good straight cut, especially those cuts that you can't reach with your tape and knife. Figure 13–4 shows how to properly position the T-square. Put a hand on the top, your knee in the middle, and the toe of your boot at the bottom of the square. Make the cut from the top of the sheet down to the middle, then from the bottom up to the middle, completing the cut. This way you won't prematurely dull your blade by running it into the floor.

Now that you've cut a straight line, what's next? Snapping the sheet. Lean the sheet out, reach over it and smack it once, good and hard, right along the cut. Now "score" the paper on the back of the sheet. Do this by reaching over the sheet and cutting along the fold, or by folding the sheet back and cutting through the open fold. A 24- to 48-inch "back-cut" is usually plenty. To finish it up, fold the sheet back a little to get some momentum going and quickly fold the sheet back, snapping the sheet along the cut. Look at Figure 13–5. Keep your knee placed close to the cut to help control the sheet as it's being snapped. This works well, but be careful not to get the skin of your knee or leg in the cut. You don't want to get pinched this bad!

Another common cut is the 45-degree, or bevel cut, on the edge of a sheet to form angles or slide behind door and window jambs. This is a particularly dangerous cut if you aren't careful. Use a sharp blade and pay close attention to what you're doing. Figure 13–6 shows a sheet properly beveled to slide into a door jamb.

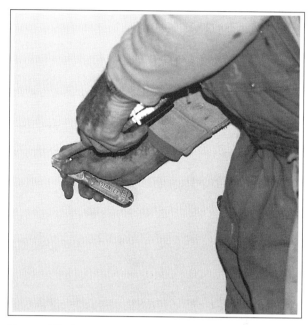

Figure 13–3. Hold the blade of a utility knife to the end of the tape to cut a straight line.

Figure 13–4. Use a T-square to make a square cut across the width of the sheet. You'll find many other situations where a 48-inch square is handy while framing.

Cutting with a Router

A drywall router is about the same as any other router, without the handles and all the fancy attachments (Figure 13-7). These drywall routers, combined with bits specifically designed for cutting drywall, are an industry standard. Use the router to cut out light boxes, recessed lights, patterns and just about anything else you'd otherwise cut with a hand saw (keyhole saw). They're also a big help when cutting gypsum coreboard, and will save you countless manhours when "topping out" the drywall above ceiling height.

Here's the procedure for cutting out electrical boxes and recessed lights. Your first step is to lay out the center of the box on the sheet. You do this by measuring from either the top or bottom and one end of the sheet, depending on what the two edges will butt up against. Use your axe handle to make sure all the wires are stuffed well back inside the box. This protects the wires from the router bit. Then tack the sheet in place. Use as many screws as you feel you need to hold the sheet in place, but don't put any screws closer than 24 inches to the box. The pressure caused by screwing too close to the box will blow out a side of the box as you're cutting it.

Now you're ready to cut it out. Make your first few cuts slowly until you get a feel for it. First, turn the router on and start the cut by inserting the bit through the center mark. Cut from the center mark to the left side of the box until you feel the router bit come into contact with the lip of the box (mud ring). Then *slowly* pull the bit out just enough to allow the bit to slide over to the outside of the mud ring. Finish up by cutting counterclockwise around the mud ring. Just guide the router and let it do the work. Never

Figure 13-5. With the first sheet scored and back-cut, set your knee just above the cut and snap the sheet sharply back to you.

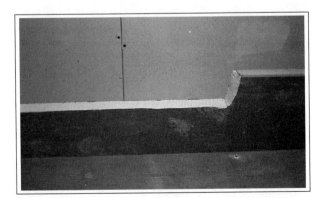

Figure 13-6. The back side of a sheet that has been cut to break over a door jamb. The edge of the sheet that will slide behind the jamb is beveled on the back side of the sheet, making it easy to slide it in behind the jamb.

Figure 13-7. Score the curve of the radius with a razor knife, then use a router to cut out a radius pattern. Cut counterclockwise and the router bit follows the line you scored in the drywall.

force it or you'll break the bit. Figure 13-8 shows a properly-cut box. You'll be whipping these out in a heartbeat before you know it. Once the cuts are complete, stand aside and hit the sheet close to the cut to knock the dust out of it. This is especially important in ceiling work. Screw the sheet off and you're ready to move on.

Setting an Outside Corner with a Rip

In metal stud framing, one of the first things you probably noticed (especially you wood framers) is the flimsiness of the light gauge metal studs. Much of the light gauge stud's rigidity comes from the drywall. This is also true with outside corners. That's why you set the corners as the walls are rocked.

Cut the rock right to the edge of the stud, as in Figure 13-9. Screw off the hard side (back) of the corner stud completely. You won't screw off the other side of the corner stud until the corner is set. As you rock the opposite side of the corner, you may find it best to run the rock wild past the corner. For this example, however, we're cutting the sheet to fit. With both sides of the corner rocked, find or cut a rip with a good bound (recessed) edge and use it to set against the corner. The side of the corner with the hard side of the corner stud is where you'll check it. Run a screw into the hard side of the corner stud, but leave it sticking out enough to grab it with your end nippers. You can use that to pull any dips out of the corner, as shown in Figure 13-10. As you get the corner adjusted to perfection, screw off the second side of the corner to hold it solid.

Screwing Off the Drywall

One of the most important factors in hanging the drywall is screwing it off properly. In most parts of the country, the drywall work is inspected just like the framing. The inspectors will focus on the spacing of the screws and their pattern. They'll also look at depth of the screws. The paper is the strength of the drywall. If you

Figure 13-8. Routers are faster and cleaner at cutting drywall. This box is typical of the quality of work that you can accomplish with a router.

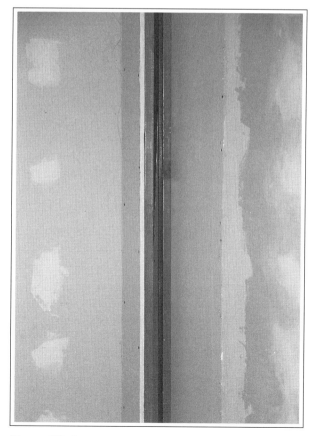

Figure 13-9. Drywall installed on one side of the column, with the sheet cut right to the edge of the stud.

Figure 13–10. Use a pair of end nippers to grab a drywall screw and pull a dip out of an outside corner. Then screw off the opposite side while you set the corner.

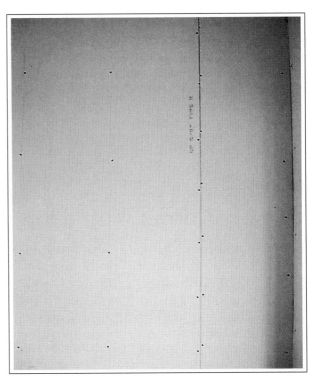

Figure 13–11. A sheet "stood up" and screwed off every 8 inches along the seam and every 12 inches in the field.

sink a screw so deep that it tears the paper, it's not holding anything. But the head of the screw *must* be sunk below the surface of the sheet. According to most of the tapers I've known, leaving a screw hanging out ranks just one step above kicking your dog! Here's a handy tip I learned on the job. If you're in doubt about a screw being deep enough, run the edge of your tape measure case over the screw. If the screw isn't deep enough, you'll hear it.

The pattern and spacing of the screws will vary. I've done shear walls on the West Coast where the spacing was 2 inches between screws. On the other end of the spectrum, the first layers of double- and triple-layer walls are simply tacked up around the perimeter with one screw in the field. The most common pattern I've found around the country is 8 inches on the perimeter (outside edge), and 12 inches in the field (middle of the sheet). This pattern will work for sheets either stood up (Figure 13–11) or laid down. Notice that in both of these figures the screw patterns are neat and consistently even. You don't have to measure the spacing between the screws. You'll just learn to eyeball the correct spacing.

It's acceptable practice to lay out the studs on the sheets. You can see the layout lines on the stand-up sheet in Figure 13–11. Laying out the studs not only eliminates misses in a sheet (which is grounds for dismissal at some outfits if you do it too often) but also keeps the studs straight. Keep in mind that for these layout marks to work out, the sheets must be plumb or level. Once you've set the first sheet of a wall, make sure the rest are in line by keeping the joints tight.

If you've never run a screw gun before, take some time to practice with it before you tackle the drywall. You can lock the screw gun on, using the lock that's located under the trigger on many models. The shaft of the screw gun is magnetic to hold the screws to the bit tip. Both are housed in an adjustable nose cone that sets the depth of the screws. Screw guns are also reversible, which comes in handy, though you should never change the direction of the gun while it's running.

Develop your own style of "screwing off" that works for you. The basics should work something like this. Keep the gun running and 10 to 15 screws in your free hand. As you run one screw into the wall, your fingers are getting the next screw ready to put on the bit tip. Hold the screw gun in your hand by the back of the motor casing and run the trigger with your ring finger. Holding it by the pistol grip causes a lot of stress on your wrist. Like many tricks we've discussed, this one's pretty simple. It just takes some practice to master.

Tying in Slap Studs

The slap studs (also called *sliders* or *slammers*) are a main source of stability for the light gauge metal stud walls, and they're all that make an inside corner solid. That makes properly tying in the slap studs an extremely important step in the rocking phase of the job. The rock will often be slid through, past a wall as a solid sheet. You can also split the rock at the center of the slider. In either case, tie the slider in, using one of the following methods:

1. Back-screw the slider from the opposite side of the wall. Screw off the slider to the drywall from the brown side of the sheet. Space the screws approximately 16 inches apart, as shown in Figure 13-12.
2. Tie in the slider by screwing through the slider into the drywall. Run the screws in at an angle, with the screws alternating side to side (Figure 13-13) or up and down (Figure 13-14). The opposing angle of the screws will hold the slider tight to the drywall, forming a solid corner. If the drywall is split behind the slider, tie both sides to the slap stud, regardless of which method you use.

That wraps up our discussion of drywall methods, and finishes this book. Besides the how-to information, there's one thing I want

Figure 13-12. Two stand-up sheets butt together on the slider and both have been back-screwed to the slider.

Figure 13-13. The leg of this stud has been cut away to make it easier to see the screws run through the stud into the drywall at opposing side-to-side angles.

Commercial Metal Stud Framing

you to take away with you. Remember what I said: "Quality first. Quantity comes with experience." As you begin watching others cut corners to go faster, *before* you adopt any of their methods, keep an eye out to see if any of their work has to be reworked. You don't need to adopt bad habits. Take pride in your work. This is more than just a trade — it's a craft.

Quality craftsmanship combined with good work ethics will soon show up on your paycheck.

And there's one more thing. Enjoy yourself.

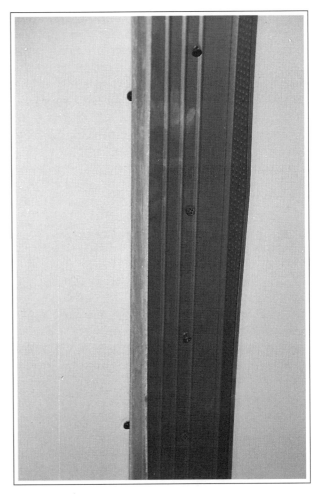

Figure 13–14. The stud is tied in with the screws running through the stud into the drywall at opposing up-and-down angles.

GLOSSARY

A

Accordion plate A deep leg deflection plate that is deeply creased along both legs, allowing the plate itself to expand and contract without removing the framing screws that hold the studs on layout in the plate.

Atrium A large open-air cavity that runs from the ground floor to the roof of a project.

Attenuation Controlling or restricting the transmission of sound from one room to another. The term attenuation is commonly associated in the trade with insulation — for example *sound attenuation batt* or *sound batt,* as it's commonly called.

B

Back-cut A sheet of drywall is scored and snapped at the desired length, then the second cut (the "back-cut") is made along the fold on the back side of the sheet.

Banjo A tool used by the drywall finishers that, when loaded with drywall joint compound and a roll of drywall tape, dispenses the tape precoated with joint compound.

Bar joist The webbed cross-bracing spanning horizontal weight-bearing girders of the red iron superstructure that carry the floor or roof.

Bastard joint A drywall joint with either one or both of the ends of the sheets forming the joint having been cut on the job (not at the factory).

Bearing wall A wall that carries all or part of the load (weight) of the floor or roof above.

Bit tip Number 1 and 2 Phillips bit tips are used in screw guns to drive the drywall and framing screws in place. The bit tips snap into a magnetic shaft that helps hold the screws on the bit tip.

Blistering The appearance of bubbles on the surface of the drywall as it's painted. The blisters occur when the face paper of the drywall is torn loose. This problem is fixed by tearing off the loose paper.

Box beam header A structural weight-bearing header that's built from two tabbed pieces of stud and two pieces of plate. The box beam header carries the weight over the rough opening of doors and windows, just as a bulkhead does on a wood-framed job.

Break stud The stud on which the edge or end of the sheets of drywall butt together. On a wall that's properly laid out, the edge or end of the sheets will hit exactly in the middle of the break studs.

Bump it down Depressing the barrel of a shotgun (powder-actuated nail set) in the firing position and releasing it, which lessens the force at which the pin is driven. The more times the shotgun is bumped down, the less the force. This technique is also called *milk it down.*

Butt joint Fitting the factory ends of two sheets of drywall together on a break stud when laying down the drywall. In contrast, when either one or both of the ends of the sheets forming the joint have been cut (not factory), it's known as a *bastard joint*.

C

C-clamps Spring-loaded locking clamps with large 4- or 6-inch jaws that are used to clamp the metal stud material together as it's fastened in place.

Chase wall Two walls built back-to-back with a space in between them to hide plumbing lines and/or duct work. The demising walls between restrooms on most commercial projects are chase walls, allowing the restrooms to share the same water feed and drain main lines.

Chop saw A powerful electric saw mounted on a heavy-duty base that uses a 14-inch abrasive blade to cut the metal stud framing materials, often by the bundle.

Cold-rolled channel Galvanized steel channel (16 gauge, $1/4$-, $3/4$-, $1 1/2$- and 2-inch) used to reinforce structural metal stud walls as well as the chief source of support for suspended drywall ceilings. The cold-rolled channel is commonly referred to as *CRC*.

Computer floor A raised flooring system of 2- x 2-foot weight-bearing floor tile supported by adjustable stands that are set to a laser beam. This system leaves a crawl space for large conduit and miles of wiring that's accessible through the removable tile. When working in an area with a computer floor, all elevations must be figured from the computer floor, *not* the concrete slab.

Control joint A $3/4$-inch gap in the framing and drywall of walls and ceilings that relieves the stress caused by the expansion and contraction that's natural in all buildings.

Coreboard Gypsum drywall panels (1 inch thick) used to build fire-rated shaft liners and shaft walls.

Cripples The studs that run from the plate to headers above and below rough openings in the framing.

Cross bracing Two-inch, 18-gauge strapping welded to structural studs, forming a giant X, to prevent the wall from racking.

Crow's foot An arrow mark that's commonly used to mark the exact point of wall lines and other layout points.

Cut A detailed drawing in the specifications section of the prints that gives a close-up view of the work.

D

Deck When framing walls up to the floor or roof above, the underside of the floor or roof is referred to as the deck.

Deep leg plate A deflection plate with legs that are 1, 1$^{1}/_{2}$ or 3 inches deep. It allows the studs in the wall to be cut short so that the deep leg plate and the deck it's shot to can raise and lower on the studs with the expansion and contraction of the deck or roof above. For use as top plate only.

Deflection A movement in a floor or roof with the addition or subtraction of weight (people and furnishings on a floor, snow and ice on a roof).

Demising wall A wall separating two areas. Also called a *party* or *parting* wall.

Door clips Eighteen-gauge metal inserts that are installed inside the throat of hollow metal door and window jambs, allowing the jambs to be fastened to the metal stud framing. The clips are sometimes welded into the jambs by the manufacturer, but usually they're simply snapped in place by the carpenters in the field.

Door studs These studs are the equivalent of the trimmer studs in wood framing, but the door studs run continuously from the bottom to the top plates. The door studs will be at least 20 gauge and are screwed off to the door clips to secure the jamb in the wall.

Double egress doors These are 4'-0 and 6'-0 door jambs that will have two doors hung from them. This type of door and jamb is very common in hospital construction.

Draft stop Fire-rated drywall installed on the trusses in the plenum (void) above a ceiling to slow or stop the free flow of air through the attic space.

Drop Another term used to describe a *soffit*.

Dryline A sturdy braided string line used to straighten the framing of walls, ceilings, and soffits.

Drywall channel A furring material that's used to fur out walls and columns, and is one of the primary framing components in a suspended drywall ceiling. It's also commonly called *DWC* or *hat channel*.

E

Ears A cut made to the end or ends of a piece of plate when cutting a header. Cutting the ears to the plate involves cutting away the hard side (back side) of the plate, leaving the legs or "ears" of the plate, which will be screwed off to the studs the header is spanning.

Elevations The heights of all the work on a construction project, such as the ceilings, soffits and countertops.

End nippers Broad-faced wire-cutting pliers used to cinch up and cut tie wire. These are must-have tools in suspended drywall ceiling work.

Exterior gyp A very dense, dark-papered gypsum drywall panel that's specifically designed to withstand the elements on the exterior of commercial buildings. These gypsum panels are available in both 4- x 8-foot and 2- x 8-foot sheets, with either plain or tongue-and-groove edges.

Extra work order A written guarantee of payment for work done that was not included in the blueprints or bid package. Extra work orders cover conditions missed by the architect and problems caused by work done by the General Contractor or other trades. This work is referred to as T&M (time & material) because it's charged by the total manhours and materials needed to do the job.

F

Finish floor The floor coverings, such as tile and carpet, that will cover the concrete slab. All the interior elevations are given off of the finish floor, and will be noted in the prints as "off F.F."

Finishers The drywall finishers, or *tapers* as they're also known, conceal all the joints, corners and raw edges of the drywall with drywall tape and joint compound *(mud)*.

Fire tape The process of sealing the joints and raw edges in the drywall above the ceiling height with drywall tape and joint compound to control the spread of fire and smoke.

Foil-back rock Gypsum drywall panels with a thin film of reflective foil laminated to the backs of them. When hanging foil-back rock, you have to be careful to avoid tearing the foil off the drywall.

G

Galvanizing A zinc coating placed on steel materials, including metal studs, that protects them from corrosion.

Gang box Large metal tool box that's used to lock up an entire crew's tools.

Gas-actuated nail set A power nail driver used to fasten the metal stud materials to concrete and masonry. This power fastening tool uses propane gas ignited by a battery-generated spark to drive a piston, which sets the pins.

General Contractor The person ultimately responsible for the work done by all the trades. The metal stud framers, plumbers and electricians are all subcontractors hired by the General Contractor.

G.F.I. Ground fault interrupter, an in-line circuit breaker added to the extension cords used to power all power tools. G.F.I. receptacles are required by OSHA for all power tools.

Girders The main support beams that carry the floor and roof trusses.

Gospel line See *Reference line*.

Gravy work The easy work on a job: work with no hassles and no obstacles.

Grid ceiling A suspended ceiling system using 1-inch-wide mains and tees to support 2- x 2-foot and 2- x 4-foot acoustical ceiling tile. This ceiling system is also commonly called an *acoustical* or *dropped* ceiling.

Gusset A brace cut from metal stud material that's used to support furring walls and other framing conditions.

H

Hanger wires Eight-gauge galvanized suspension wires used to support the cold-rolled channel and in turn the rest of a suspended drywall ceiling.

Hard number An exact measurement, with no allowance for play.

Hard side The back or solid side of a metal stud.

Hatchet A drywaller's hammer with an axe edge on one side and a dimple-faced hammer head on the other.

Hi-lo plate This plate, with a 1-inch-tall front leg and a 2-inch back leg, is used for the top and bottom plate as well as building the corners when framing shaft walls. The short front leg of the hi-lo plate allows for easy installation of the heavy coreboard panels.

I

I-beam The I-beams get their name from their shape. They're the main structural supports of the red iron superstructure, carrying the load of a building over large spans. I-beams are used for both vertical and horizontal support.

In between This term is used when giving and receiving numbers for cutting headers. "In between" refers to the length of the header in between the shoes.

Inspection The process of a city or county inspector or a civil engineer checking the work being done on a project to ensure it meets all applicable building codes.

J

Jamb Hollow metal door and window jamb used on commercial projects that is prefabricated with all three typical wood jamb components (the jamb, stop and trim) combined.

Jamb anchors Cleats welded inside the throat of a door jamb at the bottom, used to shoot or tap-con the jamb to the concrete floor.

Jig A pattern in which small pieces of metal stud are installed to prefab several identical parts of an intricate soffit. The completed parts are referred to as *jigs*.

J-metal A trim metal used to cover the raw edge of the drywall when it butts up to another material such as brick. The J-metal #200-A trim metal is shaped like a J and is slipped over the raw edge of the drywall as it's hung.

Journeyman A tradesman with enough experience or training in an apprenticeship program to complete any given task with little or no supervision; a designation of competence.

K

Kicker A metal stud (usually $3^{5}/_{8}$ inch) cut with a shoe on each end that's used to brace wall, ceiling and soffit framing. Framing braced with a kicker is referred to as *"kicked off."*

King studs The studs installed tightly up against the door/window studs when framing in a jamb or rough opening.

L

Laminate To bond materials together using adhesives.

Laser A precision leveling and plumbing tool that emits a powerful concentrated beam of light that can be read easily over great distances.

Lath screws Self-drilling framing screws used to fasten the studs and plate together. Also commonly called *S-12s*.

Layout The numerical spacing of the framing according to the prints as well as the spacing of the studs and other framing components.

Lead board Sheets of drywall with thin sheets of lead laminated to the back. This type of drywall is used to block the radiation emitted by x-ray equipment in medical buildings.

Lift Short for *scissors lift*. A gas or electric driveable scaffold that is common to the trade.

Lineman's pliers Heavy-duty bevel-nose pliers with a strong wire cutter that easily cuts heavy-gauge suspension wire. These pliers are a must in suspended ceiling work.

L-metal Like J-metal, L-metal (#200-B) is a thin trim metal used to cover the raw edges of drywall when it butts to another material. The L-metal is applied after the drywall is hung, requiring a 1/4-inch gap be left between the rock and the material it butts up against.

Lookouts Short pieces of stud that are cut and shot to the red iron to support the framework of soffits when the framework encases the red iron.

Lull A large four-wheel-drive all-terrain forklift that's common on many construction projects.

M

Mains The chief components in a grid ceiling system. The 12-foot vinyl-coated mains are suspended with hanger wires and, in turn, support the weight of the rest of the ceiling.

Manhour A labor unit representing one worker working for one hour.

Manhours The time it takes to do a job multiplied by the number of carpenters doing the work (4 carpenters x 2 hours = 8 manhours).

Married to it When you're forced to build a wall off layout or out of plumb because another trade's materials don't follow layout. You're "married to" the condition.

Mechanic A journeyman framer who has mastered the trade.

Mechanical fastener Fasteners such as lead anchors and tap-con screws that must be worked into the concrete or steel to fasten the framing in place.

Milk it down Depressing the barrel of a shotgun (powder-actuated nail set) in the firing position and releasing it, which lessens the force the pin is driven at. The more times the shotgun is bumped (milked) down, the less the force. This technique is also called *bump it down*.

Monocoat A thick fire-retardant insulation that's mixed with water and sprayed on the red iron from a spray rig or hopper. Once sprayed in place and dried, the monocoat protects the red iron superstructure from the intense heat of a fire, which could cause the steel to fail and the structure to collapse.

Mullion The framework which carries several large panes of glass that create a glass wall.

N

Non-com Short for *non-combustible*, referring to materials that resist fire.

Nut driver A hex head driver that snaps into the shaft of a screw gun just like a bit tip, used to drive hex head fasteners like tap-con anchors.

O

On center The spacing of the studs in a wall, as well as other materials, measured from the center of one member to the center of the next member. The prints abbreviate "on center" as *O.C.*

Overs and ups A method of giving measurements (numbers) for obstacles in the wall when hanging drywall. The *overs* refer to the horizontal numbers and the *ups* refer to the vertical numbers.

P

Pad The flat, level steel-reinforced concrete base of a building, also called a *slab*.

Panheads Sharp-point framing screws used to fasten light gauge framing materials together.

Parapet wall A wall that runs up above the roof line.

Perimeter The distance around the outside.

Pins The nails driven into concrete and steel with a gas- or powder-actuated nail set.

Plate The track or runner that is the top and bottom framing member of a metal stud wall. Also, the material headers are cut from.

Plumb bob A small cone-shaped weight made of brass or steel that, when suspended by a braided string line from the deck to the floor, will read exactly straight up and down (plumb).

Pony clamp A small, spring-loaded clamp.

Pony wall A very short wall, also called a *knee wall*.

Powder actuated Powered by gunpowder, as in a powder-actuated nail set, which uses a .22 caliber load to drive special nails (or pins) into concrete and steel.

Punch list A list of repair work, mistakes and omissions that is issued to each trade near the completion of a project.

Q

Quickie Saw A chain saw fitted with a 14-inch abrasive metal cutting blade, used for cutting extremely wide and heavy gauge studs.

R

Radius plate Plate with relief cuts spaced 1 inch apart that allow the plate to form curves or arches. Also plate formed in a radius plate bender.

Railroaded joints Drywall joints where the sheets break on the same stud. This creates a weak spot in the wall and a joint that's hard for the tapers to finish.

Rake wall The top of this type of wall has a sloping top plate.

Rated wall A wall built to control the spread of fire, commonly called a *fire wall*. A rated wall will be framed and rocked to the deck, then fire taped above ceiling height. These walls will often require multiple layers of drywall with the joints between each layer staggered and often fire taped.

Red iron The I-beams and girders that make up the superstructure of a large commercial project. The "red" comes from the red oxide paint the steel is painted with.

Reference line The true center line of a building snapped out on the floor, commonly called the *gospel line*. The reference lines are used to establish the wall lines when laying out an area.

Resilient channel A thin furring channel that's screwed off horizontally to wood and metal stud framing to control sound, also called *R.C. channel*.

Right angle An angle that's 90 degrees, or square.

Right angle drill motor A drill motor with an angled head fitted with a bit tip used to get screws into tight spots, also called an *offset drill*.

Rigid X ceiling system A suspended drywall ceiling system framed with grid ceiling style materials.

Rip A lengthwise strip of a sheet of drywall.

Rough opening The opening in the framing for door and window jambs. Abbreviated in the prints as *R.O.*

Run it wild Letting the end of a stick of plate or other material run out past an actual corner point, and cutting it to length as the other side of the corner is worked.

S

Saddle tie The knot used to tie suspension wires to the cold-rolled channel.

Scab A piece of plate used to splice two short pieces of stud together to make one long stud, by scabbing it together.

Score it Cutting a groove in either drywall or stud material with a utility knife so it's easier to bend or snap.

Screw gun A drill motor tool with a clutch and adjustable depth-setting nose cone used to drive framing and drywall screws.

Shaft wall A coreboard (1-inch-thick drywall) wall used to line elevator shafts, stairwells and large ductwork trunk lines to provide fire protection.

Shoe Cutting through the legs on the end of a stick of plate or stud so that the end will fold back to form a shoe. A shoe is typically 3 inches long and is commonly used in cutting headers and kickers.

Shotgun Slang name for a powder-actuated nail set used to shoot the metal stud materials to concrete and steel.

Slider The stud at an inside corner of two intersecting walls that is left loose, allowing the drywall on one wall to slide though the framing of the second. Once the drywall slides through, the slider is tied into the rock. This saves two studs per corner compared to the same corner framed with wood studs. Also called a *slap stud*.

Soffit A covering extending from the top of cabinets to the ceiling, or over the space under the eaves of a building.

Soft side The open side of a metal stud.

Splay wire A suspension wire run down at an angle past an obstacle.

Splice To join two short pieces of material together to make one long piece.

Spreaders Thin pieces of metal channel tack-welded to the bottoms of door jambs to hold them at the proper width during shipping and installation. Once the door jambs are shot down, the spreaders are knocked out with a chisel.

Stiffback A stud set on edge on top of the ceiling studs of a small ceiling and screwed off to them to help carry the weight of the drywall.

Straightedge A drywall rip, a level, or a metal stud used to transfer a wall line or other point.

Stringer A large aluminum scaffolding walk board.

Superintendent The person on the job who's responsible for the entire project and makes all major decisions. If something goes seriously wrong, he gets the heat. Then he'll come after you!

Superstructure The red iron frame of a building consisting of I-beam girders and piers which support the entire structure.

T

Tab A method used to fasten metal stud materials together. The legs of the stud or plate are cut off, about 1 inch back from the end, leaving a tab which overlaps the adjoining material providing a fastening point.

Tap-con A system used in a hammer drill consisting of a concrete drill bit and a shaft housing a nut driver that slips over the bit. Used to drill a hole in concrete and drive a fastener into it.

Three-four-five A common method of squaring an intersecting wall line off of a wall line already snapped, using only a tape measure, pencil and chalk line.

Tie in Screwing off a slider (or slap stud) to the drywall of an intersecting wall once the rock is slid in place.

Tie wire Galvanized 18-gauge wire used to tie the drywall channel to the cold-rolled channel of a suspended drywall ceiling system.

Time & materials The method of calculating how extra work orders are charged to the General Contractor, with time being the total number of man-hours plus the materials used.

Topping out As the walls are rocked, the top edge of the drywall will usually only extend 4 inches above ceiling height. On fire, smoke and sound walls, the drywall will continue up from this point to the deck of the building.

Trapeze A short piece of cold-rolled channel suspended by two hanger wires to span an obstacle. A single hanger wire is then dropped from the cold-rolled channel, getting the wire directly over where it's needed.

Trusses Lightweight members of the red iron superstructure that span the main horizontal girders to carry the load of the roof or floor above.

Type X drywall Burn-resistant fire-rated drywall.

Typical A framing condition that recurs throughout the job is referred to in the prints as typical.

U

U.S.G. Handbook United States Gypsum company handbook, one manufacturer's guide to material specifications and procedures.

Utility angle A 90-degree angle metal with flange widths ranging from $7/8$ to $2^{1}/_{2}$ inches wide, used in many framing situations. The most common use is forming the perimeter of a suspended drywall ceiling.

V

Vapor barrier Plastic sheeting that is glued or screwed to the metal studs before the walls are rocked. Its function is to prevent drafts, moisture penetration and condensation.

W

Walk-up An aluminum workbench that has extendable legs and folds neatly up for storage, commonly used in the trade.

Wall legend The section of the prints that indicates (usually by number) the specifications of each wall type on the project.

Water level A 50-foot length of clear plastic tubing attached to a small water reservoir, used to establish and maintain a constant elevation throughout a work area.

Window studs These studs are the equivalent of the trimmer studs in wood framing, but they run continuously from the bottom to the top plates.

Wire tier A tool that mounts to the end of a piece of conduit, used to tie suspension wires to the tops of the bar joist.

Wrap it To cover I-beams, girders and other members of the superstructure tightly with drywall to protect it from fire. This term also refers to rocking the inside of rough openings.

X Y Z

Z-furring channel Z-shaped metal furring strips with one flange that's shot to concrete and masonry, and another flange that holds Styrofoam insulation in place.

INDEX

A

Accessibility for the Disabled Act 6
Accordion plate 27-28
Adjustable clip 96
American-made tools, using 50
Angle iron 121
    welding studs to 123
Arc, striking 69
Arches
    cutting archway 92
    forming radius 93
    framing 28, 91
    gyp board, installing 91
    plating radius 92
    striking an arc 69

B

Back blocking joints 136
Back-cut, drywall 185
Backing out joints 136
Bar joist
    bridging 47
    plating to 16-17
    tying hanger wires to 44-45
    wiring DWC 139
Basic skills, metal framing 17
Benchmarks
    establishing 42-43
    shooting in 121
    soffit 56
Bend point 48-49
Bending suspension wires 48-49, 172
Black iron 46
    lacing with 118
Black magic markers 121
Blueprints
    locating soffit dimensions 55, 151
    reading 3
Bottom plate 13-14
    columns 142, 147
    corners, forming 79-80
    distance between pins 78
    installing heavy gauge 116
    plumbing soffit corners to 59
    shooting down 78-80
    simple framed soffit 160-161
    skylight 182
Bottom soffit studs, plating 59, 61-62
Bound edge, drywall 187
Box beam header 91, 119
    building 37-38
    door jamb 101
Braces
    gussets 31
    kickers 30
    simple stud brace 32
Bracing
    studs 110
    unequal walls 20
Bracing, diagonal (cross) 127
Bridging
    bar joist 47
    corrugated metal decking 16
Building methods, steel framing 75-76
Building with jigs 63-66
Bulkheads 170
Bullnose cutters, heavy gauge 11
Butt joints, coreboard 135

C

C-clamp 122
C-H studs 130, 134
Canopy soffits 69
Caps, parapet walls 124
Casework 151
Ceiling
    dialing in 175-176
    elevation, finding for suspended
        ceiling 175-176
    framing 42
    seismic 45
    stiffbacks 39
Ceiling systems 171
    hard lid 177
    lay-in 176-177
    suspended drywall 172
    Rigid X 176
Certified welder 123
Chalk lines
    bottom plate, setting 61-62
    snapping across long distances 139
    top plate, setting 125
    wall layout 6, 78
Channel thicknesses, DWC 106
Chase walls 14, 90
    framing around obstacles 91
Chop saw, cutting structural studs 116
Clamps 121-122
Clean work area, importance of 78
Clip with straps 96
Clips 96
    deflection 116
    slide 121
Coffered ceiling 162
    forming 155
Cold-rolled channel (CRC) 23
    at wall expansions 33
    lacing structural deflection wall
        122
    lacing studs for structural walls 118
    supporting with suspension wire
        46, 49
    suspended drywall ceilings 45, 173
    suspension wall studs 168
    tack welding to studs 23, 119
    tying at corner joints 53
Color-coding, stud gauge 115-116
Columns
    bottom plate 142
    concrete, framing around 109
    corner guards 144
    drywall, framed 146-149
    drywall rips 148
    establishing frame line 142
    framing 141
    framing cornice 144-145
    lookouts 142
    speed-framed 146-149
    three-sided 146-149
Computer floors 43
    door jambs, installation 100
Concrete
    columns, framing around 109
    decks 15
    piers, column 142
    truss systems 15-16
Concrete pins 118
Conduit rack, headering 37
Connecting framing members 29
Control stud, horizontal 59
Coreboard 131
    backing joints 135
    staggering joints 135-136, 137
    standing 133-135
Corner guards, column 144
Corner jigs, setting 163
Corner soffit studs, plumbing 58-59
Corners, blowing out 108
Corners, forming 14-15, 17-20
    bottom plate 20, 79-80
    CRC 53
    fire-rated walls 137
    freestanding walls 88-89
    hard 116
    intersecting walls 18-19
    parapet walls 126
    setting with a rip 187
    shaft wall 132-133, 135
    shooting to DWC 108
    top plate outside 18
    tying to preexisting condition 20
    unequal walls 18-19
    utility angle 44
Cornice, framing for column 144-145
Cornice studs 146
Corrugated metal decking 16
Crew requirements, plating 14
Cripples
    double cutting for CRC 120
    framing 85
    installing door jamb 99
    installing window jambs 104
Cross bracing 127
Cross channel 177

Index

Cross tees 177
Crown cornice, stair step 144-145
Cutting
   archway 92
   box beam header 37-38
   coreboard 131
   door studs 98
   drywall 184-186
   ears 30
   ears for header 36
   hand 10
   radius plates 28
   rips 138
   shoes 29
   shoes for header 35
   soffit patterns 69
   structural studs 116
   studs, walls to the deck 82-83
   tabs 30
   templates 70

D

Deck
   concrete 15
   framing soffit to 57
   metal 16
   plating 15
Deck punch 45
Decking screws, fastening with 16
Deep leg plate 26-27
   framing shaft walls 130
Deflection clips 116
Deflection plate 116
   accordion 27
   deep leg 26
Deflection wall, structural 121
   lacing with CRC 122
Detail drawings 3
   symbols 3-4
Diagonal braces
   cross bracing 127
   unequal walls 20
Dialing in ceiling 175-186
Door headers 36
Door jambs
   computer floor installation 100
   fastening 98
   headers 99
   headering underneath 100, 101
   leveling 97
   one-piece 95
   print numbers 97
   setting 97
   shimming 97-98
   stiffening with kickers 100
   suspended 101
   three-piece 95
   tight fitting 101-102
Door studs 98
Doors, entry 120
Doors, laying out 7
Double inside corner, plating 14-15
Double-duty dryline 62
Drain racks, framing around 90-91
Drawings, detail 3
Drop
   framing 55, 151
   framing with jigs 162
   kicker jigs 163
   light trough 161
   setting corner jigs 163

   to suspended ceiling 156
   utility angle 164
Dryline
   plumbing with 59, 60-61
   setting for freestanding walls 88
Drywall
   archways, installing over 91-92
   back-cut 185
   brace 110
   corners 108
   corners, fire-rated walls 137
   cutting with router 186
   cutting with utility knife 184-185
   fastening 187-188
   fire taping 142
   hanging 187
   installation 183
   marking 184
   rips, cutting 138
   scoring 185
   setting outside corners 187
   soffit support 74
   templates, soffit 69
   thickness, planning for 6
   tying in slap studs/sliders 189-190
Drywall ceiling systems 171
   framing 41
   hard lid 177
   lay-in 176-177
   suspended drywall 172-175
   Rigid X 176
Drywall channel (DWC)
   furring with 106
   installing fire lids 139
   layout 106
   shooting to wall 107
   suspended drywall ceiling 174-175
Drywall framed columns 146-149
Drywall screws, laminating 136
Ductwork, headering around 37, 84-85

E

E-stud 133
Ear protection 72
Ears, cutting 30
   on headers 36
Electrical boxes, cutting out 186-187
Elevation
   ceiling 42-43
   computer floors 100
   deflection wall 121
   hard lid ceilings 178
   parapet walls 124
   plate line for suspension wall 169
   soffit 56
   spanning speed soffit 152
   suspended ceiling 53-54
   suspended drywall ceiling 172
Elevator shaft liners 130, 131
End nippers 50, 187
End studs, shooting in place 86
Equal elevation walls, forming corners 18
Expansion joints, wall 32
Expansion plate 26
Exterior walls
   chalking 6
   furring with Z-channel 111
Exterior window openings, framing 119
Eye protection 72

F

Factory spreader 99
Fastening
   door jambs 98
   DWC 107
   framing members 29
   heavy gauge materials 116
   joints, top plate 88
   kickers, top plate 88
   plate 13
   RC channel 113
   soffit studs 59, 61-62
   studs 84
Fastening ceiling materials
   screw method 52
   tie wire method 50-51
Finished wall dimensions 6
Fire containment systems 130
Fire extinguishers, layout for 7
Fire lids 138
Fire-rated ceilings 138
   installing utility angle 140
   taping drywall 142
   tying up DWC 139
Fire-rated jambs 95
Fire-rated walls 129, 136
   framing 17
   installing safing insulation 27
   rocking corners 137
Five-siding electrical boxes 129
Floating end soffit 154
Forming corners, bottom plate 79-80
Four-step method, obstacle layout 7-8
Frame line, column 142
Framed soffit 158
   bottom plate 160-161
   inside plate 160
   kickers 159
   layout 158
   outside corner 161
   studs 159-161
   top plate 158
Framed to the deck, chase walls 90
Framing
   arches 28, 91
   around concrete columns 109
   bottom plate 13-14
   box beam headers 37-38
   chase walls 90
   columns 141
   columns with drywall 146-149
   corners 17-20, 88
   cornice to column 144-145
   cripples 85
   door jambs 96
   drywall ceiling 41
   fastening to concrete decks 16
   fire-rated ceiling 138
   fire-rated walls 17, 137-138
   freestanding walls 86
   furred walls 105, 109
   hard corners 116
   hard lid ceilings 177
   headers 35, 85
   interior walls 77
   large windows 104
   light trough 161
   lookouts 71
   metal decking 16
   multilayer walls 80
   parapet wall to red iron 126

parapet walls 123
radius soffits 69-70
radius walls 28
Rigid X system 176
shaft liner systems 130
skylights 179
smoke walls 17
soffit to deck 57
soffits 55, 151
soffits with jigs 65-66
sound walls 17
spacing for drywall 15-16
spanning gaps with spreader 19-20
spanning speed soffit 152
structural deflection walls 120
structural exterior walls 116
structural stud walls 115
structural wall rough openings 119
stud layout 21-22
stuffing studs 22-23
suspended bottom plate 59, 61-62
suspended drywall ceilings 172
suspended speed soffit 155
tools 42
top plate 15
unequal wall corners 20
using CRC 23
using deep leg plate 26-27
wall corners 14-15
wall expansions 32-33
windows 102
with jigs 56
Framing, metal, advantages 1
Freestanding walls
 bracing studs 110
 fastening top plate 86-87
 framing 20, 86
 framing corners 88
 installing jambs 89-90
 installing kickers 87-88
 setting corners 88-89
 setting ends 86
 stud layout at top plate 87
Full-framed columns 142
Furred walls 105
Furring
 inch 'n five studs 109
 RC channel 112-113
 Z-channel 111
Furring methods, DWC 106

G

Gang box 118
Gap, setting
 door jambs 98
 window jambs 103
Gas-actuated nail set 15
 assembly 107
 using 78
Gauge block 62
Gauge, metal studs 62
 structural 115
Gospel line, establishing 6
 skylight layout 179
Green-handled snips 10
Grid ceilings 4, 176
Gusset 31
 bracing top plate 110
Gyp board, installing over archways 91-92

H

H studs 130, 134
Hallways, laying out 6
Hammer drill, using 78
Hand cutting, plates and studs 10
Hanger wires
 CRC, suspension walls 168
 suspended drywall ceiling 172
 tying up 44
Hanging drywall 187
Hard corners, framing 116
Hard lid ceilings 39, 177
 framing rough openings 179
 installing stiffback 178-179
 installing studs 178
 plating 178
 reflective ceiling plan 4
Hard number 82
Harness 121
Hat channel 52, 106
 suspended drywall ceiling 174-175
Hat channel head, shotgun 106
Hat track 106
Headers
 around obstacles 36-37, 84
 bottom plate, chase walls 91
 box beam 37-38, 91
 door 36, 99-100
 framing 85
 laying out 35
 structural 36
 window 39, 103
Heavy-gauge material, splicing 12
Hi-lo plate 130
Hollow metal jambs 95
Horizontal control stud 59
 skylight 181
Horizontal welds, studs 123
Hourly wage, metal frame construction 2
HVAC shaft liners 130
HVAC systems, framing around 41

I

"In between" measurements 29, 35, 85
Inch 'n five studs
 bracing top plate 110
 furring with 109
Inside corners
 drywall 189-190
 plating 14-15
 shooting DWC 108
Inside plate, simple framed soffit 160
Installing clips 96
Installing studs 22-23, 83-84
Insulating Z-channel 111-112
Insulation
 between plates 27
 safing 27
 sound 112
Interior walls
 chalking 6
 corners 18-19
 framing 77
 intersecting walls 23-24
 types 78-79

J

J-mold 158
J-runner 130
Jambs
 door 97
 hollow metal 95
 installing in freestanding walls 89-90
 installing in lead-lined walls 96-97
 measuring for consistent elevation 7
 tight-fitted 101-102
 window 102-104
Jig, building 63-64
Jigs, soffit 162
 framing with 56
 kicker 163
 plumbing 68
 setting 66
Joining plates 11-12
Joining studs 29
Joints
 fire taping 142
 suspended ceiling 52

K

Kickers 30
 jigs 163
 reinforcing spanning soffit 154
 simple framed soffit 159
 soffit 60
 stiffening jamb with 100
 support for freestanding wall 87-88
 suspension walls 168
King studs 98
 installing tight-fitted jambs 102

L

L-metal 158
Lacing studs 118
Laminating drywall screws 136
Lanyard 121
Large jambs, window 104
Laser 43
 plumbing exterior walls 116-117
 soffit framing 57-58
 suspended ceiling installation 54
Laser card 167
Laser, pocket 152
Lath screws 44, 116
Lay-in ceiling 176-177
Laying out patterns 69-70
Layout
 around obstacles 7-8
 column 142, 147
 door 7
 drop for light trough 162
 drywall 21
 DWC 106
 fire lid, DWC 139
 hat channel 106
 recessed items 7
 shaft wall 131
 simple framed soffit 158
 skylight 179
 soffit 55-56
 soffit lookouts 72
 structural deflection wall 121
 wall 6
 window 7

Index

Layout jigs 164
Lead-lined walls
    door jambs 96-97
    window jambs 104
Left-cut snips 10
Legend, wall 3-4
Lengths, light gauge studs 77
Levels
    magnetic 97
    torpedo 146
    water 42
Lifts, working off 166-167
Light gauge studs, basic widths and lengths 77
Light trough 161
    framing with jigs 162
    installing utility angle 164
    kicker jigs 163
    layout jigs 164
    plating 164-166
    setting corner jigs 163
Lineman's pliers 176
Long studs, standing 122
Lookouts
    column 142
    column cornice 145
    soffit 71-72
Lumber prices, instability of 2

M

Magnetic level 97
Mains, suspended ceiling 177
Marking
    drywall sheets 184
    studs, gauge 115-116
    wall layout 7
Masonry
    fastenings studs to 86
    shooting DWC to 108
Material, storing/stocking 78
Measurements, "in between" 29, 35
Measuring, right angles 5
Metal decking 16
Metal framing
    advantages 1
    price stability 2
    skills required 17
    wages and benefits 2
Metal jambs, hollow 95
Metal scoring 11
Metal stud walls
    framing 77
    structural 115
Metal trim 158
Methods, framing 75-76
Minimizing waste, importance of 77
Mullions, framing to 120
Multilayer walls, framing 80
Multiple plate lines 80

N

90-degree plumb marks, using 81
Notations, print 3
Notching plates/studs 24

O

Obstacles
    framing chase walls 91
    headering around 36-37, 84-85
    notching plates or studs for 24
    planning layout around 7-8
    plating around 80
Offset plate elevation walls, forming corners 18-19
Offset plate lines 80
One-piece door jambs 95
Openings, liners for 130
Outside corners
    bottom plate 20
    setting with a rip 187
    shooting DWC 108
    simple framed soffit 161

P

Pan-head screws 13
Parapet walls
    establishing elevation 124
    framing 123
    framing to red iron 126
    intersecting walls 126
    plating the studs 124
    setting the plate 125-126
Patterns, soffit 69
Piers 116
Pins
    concrete 118
    sheared 109
    steel 118
Pipes, headering around 84-85
Pivot point, arch 92
Planning, importance of 77-78
Plate line elevation
    establishing 57
    setting 169
Plate lines
    multiple 80
    offset 80
Plates
    bottom 13-14
    cutting 10
    deep leg 26-27
    deflection 26-28
    hi-lo 130
    inside 160
    insulation between 27
    laying out 21
    notching 24-25
    radius 28
    shaft liner 130
    shooting 13
    splicing 11-12
Plating
    arch radius 92
    around obstacles 80
    bottom studs, soffit 59, 61-62
    deck 15
    hard lid ceilings 178
    inside corners 14-15
    light trough drop 164-166
    red iron 117
    studs, parapet walls 124
    suspended speed soffit 156-157
    teamwork 78
    to bar joist 16-17
Plumb bobs
    custom 121
    using 8
Plumbing
    jambs 98
    plate line, column lookouts 143
    soffit 58-59
    structural walls 116-117
    suspension walls 167
    techniques 8-9
    wall lines 81
    with dryline 60-61
Plywood spreader 99
Plywood templates, soffit 69
Pocket laser 121, 152
    plumbing with 9
Pony clamp 121
Powder-actuated nail set
    assembly 106
    cleaning 106-107
    plating with 14-15
    using 78
    with hat channel head 106
Preexisting condition, tying to 20
Prefab shaft liner corner 133
Prepainted wall mold 176
Prepping work area 78
Price stability, metal framing 2
Prices, wood products 2
Print notations 3
Print numbers, jambs 97
Protrusion framing 71
Pulling layout 22
    stud layout, offset walls 80

Q

Quality work, importance of 77
Quality, wood products 2
Quickie Saw 24
    cutting structural studs 116

R

Racking, prevention of 127
Radius plate 28
Radius plate bender 93-94
Radius soffit patterns 69-70
Razor knife 158
    scoring with 186
RC channel 112-113
Reading blueprints 3
Recessed edge 187
Recessed items, wall layout 7
Recessed lights, cutting out 186
Red iron
    building columns around 142
    forming parapet walls 126
    piers 116
    plumbing to top piece 121
    superstructure, fastening to 17
Red-handled snips 10
Reference lines, establishing 6
Reflective ceiling plans 4, 41
Reinforcing wall studs 23
Relief cuts, radius plates 28
Resilient furring channel 112-113
Rest rooms, framing chase wall 90
Right angles, establishing 4-5
Right-cut snips 10
Rigid X system 176-177
Rips 138
    bottom, suspended soffit 157
    spanning speed soffit 152
    speed-framed column 148
    vertical face, suspended soffit 156
Rock to rock screws 136

Index

Rocking corners 108
   fire-rated walls 137
Roofing material, metal 16
Room, squaring up 109
Rough openings
   framing in structural walls 119
   hard lid ceilings 179
   window 7
Router 131
   drywall 186

S

S-12 framing screws 13, 116
Saddle tie 45-46
   CRC, suspension walls 168
   tying process 49-50
Safety equipment 121
   glasses 72
Safing insulation, installing 27
Scabbing structural plate 12-13
Scabs 13
Scaffold
   multilevel 131
   scissors lift 41
Scoring 131
   drywall 185
   metal cutting 11
Scrap, using 77
Screw gun, using 188
Screw method, fastening ceiling
   materials 52
Screw pattern, drywall installation 187
Seismic ceilings 45
Self-drilling framing screws 13
Setting
   door jamb 97
   door jamb, computer floor
      installation 100-101
   freestanding wall ends 86
   parapet wall plate 125-126
   soffit bottom plate 62-63
Shaft liner system 130
   building corners 132-133, 135
   C-H studs 130, 134
   coreboard 131
   E-stud corners 133
   fire code requirements 136
   H-studs 130, 134
   liner plate 130
   stacking liners 135-136
   wall layout 131
Shaft walls 130
Sheared-off pins 109
Shimming door jamb 97
Shoes
   bracing with 19
   cutting 29
   cutting on headers 35
Shooting
   DWC 107
   end studs 86
Shooting bottom plate 13-14
   walls to the deck 78-80
Shooting top plate 15
   structural steel walls 117
   walls to the deck 82
Shotgun hat channel head 106
Side clips 116
Simple stud brace 32, 180
   bracing top plate 110
6-8-10 squaring method 6

Skills, metal framing 17
Skylights
   bottom plate 182
   bracing to red iron 180
   framing 179
   installing soffit through 71-72
   installing studs 180-181
   top plate 181
Slammers 189
Slap studs 86, 189
   establishing 14
   use in corner framing 23-24
Slide clips 121
   welding to red iron 122
Sliders 189
   establishing 14
   use in corner framing 23-24
Smoke walls, framing 17
Snapping chalk lines 6
   top plate line, walls to the deck 81
Snips, metal cutting 10-11
Soapstone 121
Soffit
   floating end 154
   framed 158
   light trough 161
   spanning speed soffit 152
   suspended speed soffit 155
Soffits
   building through skylight 71-72
   building with jigs 65-66
   drywall support 74
   establishing elevation 56
   figuring stud length 57
   forming corner 20-21
   framing 55, 151
   framing with jigs 161
   kicking off 60
   layout 55-56
   line, preventing sag 156
   locating dimensions in blueprints 55
   lookout support 71
   plumbing corner studs 58-59
   radius patterns 69-70
   setting bottom plate 62-63
   setting jigs 67-68
   supporting 71
Soft side, stud 121
Sound transfer, preventing 112
Sound walls, framing 17
Spanning speed soffit 152
   floating end 154-155
Spanning unequal walls 19-20
Specialty items, symbols 3-4
Speed-framed columns 146-149
Splay wire 45
Splicing
   joints 87
   plate 11-12, 82
   suspension wires 46
Split point 48
Spreaders
   setting door jamb 99
   spanning gaps with 19-20
Spring clamp 121
Squaring
   a room 109
   3-4-5 method 4-5
Stacking, shaft liners 135
Staggered joints, coreboard 135, 137
Stair-step crown cornice 144-145
Standard clip 96

Steel pins 17, 118
   fastening concrete columns 109
Step-by-step methods, steel framing 75
Stiffbacks 39
   hard lid ceilings 178
Stilts, working on 152
Stocking material 78
Storing material 78
Straight-cut snips 10
Straightedge, using 33-34
Strength advantage, metal studs 1
Stretching suspension wires 47-48
   suspended drywall ceiling 172
Striking an arc 69
String line, plumbing with 59, 60-61
Structural headers 36
Structural plates, splicing 12
Structural studs
   cutting 116
   notching 25
Structural walls 115
   deflection plate 120-121
   diagonal cross bracing 127
   stuffing 118
   welding to red iron 123
Stud brace
   simple 32
   using 110
Stud layout 21-22
   plumbing 81
   pulling 80
   top plate, freestanding walls 87
   walls to the deck 82
Stud walls, structural 115
Studs
   adding for strength 85
   C-H 130, 134
   column 143, 147
   cutting 10, 82-83
   door 98
   establishing soft side 121
   fastening 84
   H 130, 134
   hard lid ceiling 178
   horizontal control 59
   inch 'n five 109
   installing in light trough drop
      164-166
   installing windows 103
   king 98
   lining up holes 23
   marking gauge 115-116
   notching 24-25
   plating, suspension wall 169
   running wild 57
   setting into structural walls 118
   shaft liner 130
   simple framed soffit 159-161
   skylight 180-181
   slap 14
   spanning speed soffit 152, 155
   standing long lengths 122
   stiffback 39
   stuffing 22-23
   suspension wall 167
   walls to the deck 82-84
   welding to angle iron 123
   window 98
Stuffing studs 22-23
   framed soffit 160-161
   structural walls 118
   walls to the deck 83-84

Styrofoam insulation 111-112
Superstructure, fastening to 17
Suspended ceilings
    adding extra support 53
    establishing elevation 42-43
    fastening joints 52
    fastening with screw method 52
    fastening with tie wire method 50-51
    forming corners with CRC 53
    framing 41
    tuning in elevation 53-54
    tying suspension wires 44
    utility angle 44
Suspended doors, installing 101
Suspended drywall ceiling
    adjusting ceiling elevation 175-176
    establish elevation 172
    hanging suspension wires 172
    installing CRC 173
    installing DWC 174-175
    installing utility angle 172
Suspended framing, bottom studs 59, 61-62
Suspended speed soffit 155
    hanging vertical face rips 156
Suspension trapeze 47
Suspension walls 166
    framing corner 20-21
    kicking off 60
    lacing CRC 168
    plating studs 169
    plumbing with laser 167
    setting plate line elevation 169
    supporting CRC 168
Suspension wires
    bending 48-49
    bridging bar joist 47
    locking into place 54
    Rigid X system 176-177
    splicing 46
    stretching 47-48
    supporting CRC 23, 49
    suspended drywall ceiling 172
    tying to bar joist 44-45
Swing, door 97
Symbols, detail 3-4

T

T-square 185
Tab method, top plate fastening 16
Tabs, cutting 30
Tags, jamb 97
Tap-con system, concrete fastening 78
Tape and pencil, marking with 184
Teamwork, plating 78
Tek screws 16
Templates
    cutting 70
        radius plates 69
34s 116
3-4-5 squaring method 4-5
Three-piece door jambs 95
Three-sided column 146-149
Tie wire method, fastening ceiling materials 50-51
Tight-fitted jambs 101-102
Tissue dispensers, layout for 7
Tools, American made 50

Top out 85
Top plate
    bracing 110
    column 143
    forming outside corners 18
    freestanding walls 86, 88
    light trough drop 162
    shaft wall 132
    shooting 15
    simple framed soffit 158
    skylight 181
    snapping line 81
    structural steel walls 117
    suspension walls 167
Torch, cutting with 116
Torpedo level 146
Track, splicing 11-12
Trade tricks, metal framing 1-2
Transferring a wall line 33-34
Trapeze, suspension 47
Tricks of the trade 1-2
Trim metal 158
Tying
    metal framing 11-12
    rock to truss 138
    suspended ceilings joints 52
    suspension wires 44-49

U

Unequal walls, bracing 19-20
Utility angle
    arch 93
    fire-rated ceiling 140
    light trough 164
    suspended ceiling 44
    suspended drywall ceiling 172
Utility knife 131
    cutting drywall 184

V

Vertical face rips 156
Vertical face soffit studs 153
Vertical stud welds 123

W

Wages and benefits, metal construction 2
Walk-up 107
Wall angle 44
Wall dimensions, blueprints 6
Wall expansions 32
Wall legend 3-4
    material requirements 80
Wall lines
    chalking 78
    chalking both sides 6-7
    plumbing 81
    transferring 33-34
Wall mold, prepainted 176
Wall types 78-79
    chase 14
    fire 130
    freestanding 20
    glass and mullion 120
    to the deck 78

Walls
    corner framing 14-15
    exterior, furring with Z-channel 111
    forming corners 17-20
    furred 105
    laying out 6
    lead-lines 96-97
    rigidity, increasing 110
    stud layout for drywall 21
    suspension, kicking off 60
Walls to the deck 78
    cutting studs to length 82-83
    ending the wall 86
    headering for obstacles 84
    plumbing wall lines 81
    shooting bottom plate 78
    shooting top plate 82
    stud layout 82
    stuffing the studs 83-84
Waste, minimizing 77
Water level, using 42
Welding
    CRC 23, 119
    cross bracing 127
    slide clips 122
    structural walls to red iron 123
    studs to angle iron 123
Wheelchair accessibility 6
Widths, light gauge studs 77
Wind factor, plumbing walls 116-117
Window jambs 95
    cripples 104
    headers 36, 103
    installing studs 103
    lead-lined 104
    setting 102
Window studs 98
Windows
    large 104
    laying out 7
Wire tier 44
Wires, suspension 44
    saddle tying CRC 168
    suspended drywall ceiling 172
Wiring, drywall to truss 138
Wood prices, instability of 2
Work area
    cleaning 78
    laying out 6
Work quality, importance of 77

X

X-bracing 127
X-ray room walls 96-97

Y

Yellow-handled snips 10

Z

Z-channel
    furring with 111
    installing insulation 111-112

Practical References for Builders

Basic Engineering for Builders

If you've ever been stumped by an engineering problem on the job, yet wanted to avoid the expense of hiring a qualified engineer, you should have this book. Here you'll find engineering principles explained in non-technical language and practical methods for applying them on the job. With the help of this book you'll be able to understand engineering functions in the plans and how to meet the requirements, how to get permits issued without the help of an engineer, and anticipate requirements for concrete, steel, wood and masonry. See why you sometimes have to hire an engineer and what you can undertake yourself: surveying, concrete, lumber loads and stresses, steel, masonry, plumbing, and HVAC systems. This book is designed to help the builder save money by understanding engineering principles that you can incorporate into the jobs you bid. **400 pages, 8½ x 11, $34.00**

Basic Lumber Engineering for Builders

Beam and lumber requirements for many jobs aren't always clear, especially with changing building codes and lumber products. Most of the time you rely on your own "rules of thumb" when figuring spans or lumber engineering. This book can help you fill the gap between what you can find in the building code span tables and what you need to pay a certified engineer to do. With its large, clear illustrations and examples, this book shows you how to figure stresses for pre-engineered wood or wood structural members, how to calculate loads, and how to design your own girders, joists and beams. Included FREE with the book — an easy-to-use version of NorthBridge Software's *Wood Beam Sizing* program. **272 pages, 8½ x 11, $38.00**

Blueprint Reading for the Building Trades

How to read and understand construction documents, blueprints, and schedules. Includes layouts of structural, mechanical, HVAC and electrical drawings. Shows how to interpret sectional views, follow diagrams and schematics, and covers common problems with construction specifications. **192 pages, 5½ x 8½, $14.75**

BNI Public Works Costbook, 2000

This is the only book of its kind for public works construction. Here you'll find labor and material prices for most public works and infrastructure projects: roads and streets, utilities, street lighting, manholes, and much more. Includes manhour data and a 200-city geographic modifier chart. Includes FREE estimating software and data. **450 pages, 8½ x 11, $74.95**

Carpentry in Commercial Construction

Covers forming, framing, exteriors, interior finish, and cabinet installation in commercial buildings: how to design and build concrete forms, select lumber dimensions, what grades and species to use for a design load, how to select and install materials based on their fire rating or sound-transmission characteristics, and plan and organize a job efficiently. Loaded with illustrations, tables, charts, and diagrams. **272 pages, 5½ x 8½, $19.00**

Contractor's Guide to QuickBooks Pro 99

This user-friendly manual walks you through QuickBooks Pro's detailed setup procedure and explains step-by-step how to create a first-rate accounting system. You'll learn in days, rather than weeks, how to use QuickBooks Pro to get your contracting business organized, with simple, fast accounting procedures. On the CD included with the book you'll find a full version of QuickBooks Pro, good for 25 uses, with a QuickBooks Pro file preconfigured for a construction company (you drag it over onto your computer and plug in your own company's data). You'll also get a complete estimating program, including a database, and a job costing program that lets you export your estimates to QuickBooks Pro. It even includes many useful construction forms to use in your business. **312 pages, 8½ x 11, $42.00**
Also available: **Contractor's Guide to QuickBooks Pro *version 6*. $39.75**

Contractor's Survival Manual

How to survive hard times and succeed during the up cycles. Shows what to do when the bills can't be paid, finding money and buying time, transferring debt, and all the alternatives to bankruptcy. Explains how to build profits, avoid problems in zoning and permits, taxes, time-keeping, and payroll. Unconventional advice on how to invest in inflation, get high appraisals, trade and postpone income, and stay hip-deep in profitable work. **160 pages, 8½ x 11, $22.25**

Managing the Small Construction Business

Overcome your share of business hassles by learning how 50 small contractors handled similar problems in their businesses. Here you'll learn how they handle bidding, unit pricing, contract clauses, change orders, job-site safety, quality control, overhead and markup, managing subs, scheduling systems, cost-plus contracts, pricing small jobs, insurance repair, finding solutions to conflicts, and much more. **243 pages, 8½ x 11, $27.95**

Finish Carpenter's Manual

Everything you need to know to be a finish carpenter: assessing a job before you begin, and tricks of the trade from a master finish carpenter. Easy-to-follow instructions for installing doors and windows, ceiling treatments (including fancy beams, corbels, cornices and moldings), wall treatments (including wainscoting and sheet paneling), and the finishing touches of chair, picture, and plate rails. Specialized interior work includes cabinetry and built-ins, stair finish work, and closets. Also covers exterior trims and porches. Includes manhour tables for finish work, and hundreds of illustrations and photos. **208 pages, 8½ x 11, $22.50**

CD Estimator

If your computer has *Windows*™ and a CD-ROM drive, *CD Estimator* puts at your fingertips 85,000 construction costs for new construction, remodeling, renovation & insurance repair, electrical, plumbing, HVAC and painting. You'll also have the *National Estimator* program — a stand-alone estimating program for *Windows*™ that *Remodeling* magazine called a "computer wiz." Quarterly cost updates are available at no charge on the Internet. To help you create professional-looking estimates, the disk includes over 40 construction estimating and bidding forms in a format that's perfect for nearly any word processing or spreadsheet program for *Windows*™. And to top it off, a 70-minute interactive video teaches you how to use this CD-ROM to estimate construction costs. **CD Estimator is $68.50**

National Construction Estimator

Current building costs for residential, commercial, and industrial construction. Estimated prices for every common building material. Provides manhours, recommended crew, and gives the labor cost for installation. Includes a CD-ROM with an electronic version of the book with *National Estimator*, a stand-alone *Windows*™ estimating program, plus an interactive multimedia video that shows how to use the disk to compile construction cost estimates. **616 pages, 8½ x 11, $47.50. Revised annually**

Roof Framing

Shows how to frame any type of roof in common use today, even if you've never framed a roof before. Includes using a pocket calculator to figure any common, hip, valley, or jack rafter length in seconds. Over 400 illustrations cover every measurement and every cut on each type of roof: gable, hip, Dutch, Tudor, gambrel, shed, gazebo, and more. **480 pages, 5½ x 8½, $22.00**

Building Layout

Shows how to use a transit to locate a building correctly on the lot, plan proper grades with minimum excavation, find utility lines and easements, establish correct elevations, lay out accurate foundations, and set correct floor heights. Explains how to plan sewer connections, level a foundation that's out of level, use a story pole and batterboards, work on steep sites, and minimize excavation costs. **240 pages, 5½ x 8½, $15.00**

Residential Steel Framing Guide

Steel is stronger and lighter than wood — straight walls are guaranteed — steel framing will not wrap, shrink, split, swell, bow, or rot. Here you'll find full page schematics and details that show how steel is connected in just about all residential framing work. You won't find lengthy explanations here on how to run your business, or even how to do the work. What you will find are over 150 easy-to-read full-page details on how to construct steel-framed floors, roofs, interior and exterior walls, bridging, blocking, and reinforcing for all residential construction. Also includes recommended fasteners and their applications, and fastening schedules for attaching every type of steel framing member to steel as well as wood. **170 pages, 8^1/$_2$ x 11, $38.80**

National Electrical Estimator

This year's prices for installation of all common electrical work: conduit, wire, boxes, fixtures, switches, outlets, loadcenters, panelboards, raceway, duct, signal systems, and more. Provides material costs, manhours per unit, and total installed cost. Explains what you should know to estimate each part of an electrical system. Includes a CD-ROM with an electronic version of the book with *National Estimator*, a stand-alone *Windows*™ estimating program, plus an interactive multimedia video that shows how to use the disk to compile construction cost estimates.
544 pages, 8^1/$_2$ x 11, $47.75. Revised annually

Rough Framing Carpentry

If you'd like to make good money working outdoors as a framer, this is the book for you. Here you'll find shortcuts to laying out studs; speed cutting blocks, trimmers and plates by eye; quickly building and blocking rake walls; installing ceiling backing, ceiling joists, and truss joists; cutting and assembling hip trusses and California fills; arches and drop ceilings — all with production line procedures that save you time and help you make more money. Over 100 on-the-job photos of how to do it right and what can go wrong. **304 pages, 8^1/$_2$ x 11, $26.50**

Simplified Guide to Construction Law

Here you'll find easy-to-read, paragraphed-sized samples of how the courts have viewed common areas of disagreement — and litigation — in the building industry. You'll read about legal problems that real builders have faced, and how the court ruled. This book will tell you what you need to know about contracts, torts, fraud, misrepresentation, warranty and strict liability, construction defects, indemnity, insurance, mechanics liens, bonds and bonding, statutes of limitation, arbitration, and more. These are simplified examples that illustrate not necessarily who is right and who is wrong — but who the law has sided with. **298 pages, 5^1/$_2$ x 8^1/$_2$, $29.95**

Contracting in All 50 States

Every state has its own licensing requirements that you must meet to do business there. These are usually written exams, financial requirements, and letters of reference. This book shows how to get a building, mechanical or specialty contractor's license, qualify for DOT work, and register as an out-of-state corporation, for every state in the U.S. It lists addresses, phone numbers, application fees, requirements, where an exam is required, what's covered on the exam and how much weight each area of construction is given on the exam. You'll find just about everything you need to know in order to apply for your out-of-state license. **416 pages, 8^1/$_2$ x 11, $36.00**

Commercial Electrical Wiring

Make the transition from residential to commercial electrical work. Here are wiring methods, spec reading tips, load calculations and everything you need for making the transition to commercial work: commercial construction documents, load calculations, electric services, transformers, overcurrent protection, wiring methods, raceway, boxes and fittings, wiring devices, conductors, electric motors, relays and motor controllers, special occupancies, and safety requirements. This book is written to help any electrician break into the lucrative field of commercial electrical work.
320 pages, 8^1/$_2$ x 11, $27.50

 Craftsman Book Company
6058 Corte del Cedro
P.O. Box 6500
Carlsbad, CA 92018

☎ 24 hour order line
1-800-829-8123
Fax (760) 438-0398

Name _____

Company _____

Address _____

City/State/Zip _____
○ This is a residence
Total enclosed _____ (In California add 7.25% tax)
We pay shipping when your check covers your order in full.

In A Hurry?
We accept phone orders charged to your
○ Visa, ○ MasterCard, ○ Discover or ○ American Express

Card# _____

Exp. date _____ Initials _____
Tax Deductible: Treasury regulations make these references tax deductible when used in your work. Save the canceled check or charge card statement as your receipt.

Order online http://www.craftsman-book.com
Free on the Internet! Download any of Craftsman's estimating costbooks for a 30-day free trial! http://costbook.com

10-Day Money Back Guarantee
- ○ 34.00 Basic Engineering for Builders
- ○ 38.00 Basic Lumber Engineering for Builders
- ○ 14.75 Blueprint Reading for the Building Trades
- ○ 74.95 BNI Public Works Costbook
- ○ 15.00 Building Layout
- ○ 19.00 Carpentry in Commercial Construction
- ○ 68.50 CD Estimator
- ○ 27.50 Commercial Electrical Wiring
- ○ 36.00 Contracting in All 50 States
- ○ 42.00 Contractor's Guide to QuickBooks Pro 99
- ○ 39.75 Contractor's Guide to QuickBooks Pro *version 6*
- ○ 22.25 Contractor's Survival Manual
- ○ 22.50 Finish Carpenter's Manual
- ○ 27.95 Managing the Small Construction Business
- ○ 47.50 National Construction Estimator with FREE *National Estimator* on a CD-ROM.
- ○ 47.75 National Electrical Estimator with FREE *National Estimator* on a CD-ROM.
- ○ 38.80 Residential Steel Framing Guide
- ○ 22.00 Roof Framing
- ○ 26.50 Rough Framing Carpentry
- ○ 29.95 A Simplified Guide to Construction Law
- ○ 45.00 Commercial Metal Stud Framing
- ○ FREE Full Color Catalog

Prices subject to change without notice

Craftsman Book Company
6058 Corte del Cedro
P.O. Box 6500
Carlsbad, CA 92018

☎ 24 hour order line
1-800-829-8123
Fax (760) 438-0398

Name
Company
Address
City/State/Zip
○ This is a residence
Total enclosed_____(In California add 7.25% tax)
We pay shipping when your check covers your order in full.

In A Hurry?
We accept phone orders charged to your
○ Visa, ○ MasterCard, ○ Discover or ○ American Express

Card#_____

Exp. date_____Initials_____

Tax Deductible: Treasury regulations make these references tax deductible when used in your work. Save the canceled check or charge card statement as your receipt.

Order online http://www.craftsman-book.com
Free on the Internet! Download any of Craftsman's estimating costbooks for a 30-day free trial! http://costbook.com

10-Day Money Back Guarantee

○ 34.00 Basic Engineering for Builders
○ 38.00 Basic Lumber Engineering for Builders
○ 14.75 Blueprint Reading for the Building Trades
○ 74.95 BNI Public Works Costbook
○ 15.00 Building Layout
○ 19.00 Carpentry in Commercial Construction
○ 68.50 CD Estimator
○ 27.50 Commercial Electrical Wiring
○ 36.00 Contracting in All 50 States
○ 42.00 Contractor's Guide to QuickBooks Pro 99
○ 39.75 Contractor's Guide to QuickBooks Pro version 6
○ 22.25 Contractor's Survival Manual
○ 22.50 Finish Carpenter's Manual
○ 27.95 Managing the Small Construction Business
○ 47.50 National Construction Estimator with FREE *National Estimator* on a CD-ROM.
○ 47.75 National Electrical Estimator with FREE *National Estimator* on a CD-ROM.
○ 38.80 Residential Steel Framing Guide
○ 22.00 Roof Framing
○ 26.50 Rough Framing Carpentry
○ 29.95 A Simplified Guide to Construction Law
○ 45.00 Commercial Metal Stud Framing
○ FREE Full Color Catalog

Prices subject to change without notice

Craftsman Book Company
6058 Corte del Cedro
P.O. Box 6500
Carlsbad, CA 92018

☎ 24 hour order line
1-800-829-8123
Fax (760) 438-0398

Name
Company
Address
City/State/Zip
○ This is a residence
Total enclosed_____(In California add 7.25% tax)
We pay shipping when your check covers your order in full.

In A Hurry?
We accept phone orders charged to your
○ Visa, ○ MasterCard, ○ Discover or ○ American Express

Card#_____

Exp. date_____Initials_____

Tax Deductible: Treasury regulations make these references tax deductible when used in your work. Save the canceled check or charge card statement as your receipt.

Order online http://www.craftsman-book.com
Free on the Internet! Download any of Craftsman's estimating costbooks for a 30-day free trial! http://costbook.com

10-Day Money Back Guarantee

○ 34.00 Basic Engineering for Builders
○ 38.00 Basic Lumber Engineering for Builders
○ 14.75 Blueprint Reading for the Building Trades
○ 74.95 BNI Public Works Costbook
○ 15.00 Building Layout
○ 19.00 Carpentry in Commercial Construction
○ 68.50 CD Estimator
○ 27.50 Commercial Electrical Wiring
○ 36.00 Contracting in All 50 States
○ 42.00 Contractor's Guide to QuickBooks Pro 99
○ 39.75 Contractor's Guide to QuickBooks Pro version 6
○ 22.25 Contractor's Survival Manual
○ 22.50 Finish Carpenter's Manual
○ 27.95 Managing the Small Construction Business
○ 47.50 National Construction Estimator with FREE *National Estimator* on a CD-ROM.
○ 47.75 National Electrical Estimator with FREE *National Estimator* on a CD-ROM.
○ 38.80 Residential Steel Framing Guide
○ 22.00 Roof Framing
○ 26.50 Rough Framing Carpentry
○ 29.95 A Simplified Guide to Construction Law
○ 45.00 Commercial Metal Stud Framing
○ FREE Full Color Catalog

Prices subject to change without notice

Mail This Card Today
For a Free Full Color Catalog

Over 100 books, annual cost guides and estimating software packages at your fingertips with information that can save you time and money. Here you'll find information on carpentry, contracting, estimating, remodeling, electrical work, and plumbing.

All items come with an unconditional 10-day money-back guarantee. If they don't save you money, mail them back for a full refund.

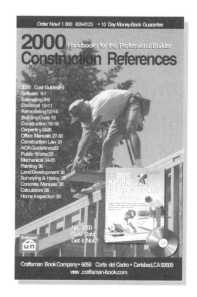

Name
Company
Address
City/State/Zip

Craftsman Book Company / 6058 Corte del Cedro / P.O. Box 6500 / Carlsbad, CA 92018

BUSINESS REPLY MAIL
FIRST CLASS MAIL PERMIT NO. 271 CARLSBAD, CA

POSTAGE WILL BE PAID BY ADDRESSEE

 Craftsman Book Company
6058 Corte del Cedro
P.O. Box 6500
Carlsbad, CA 92018-9974

NO POSTAGE
NECESSARY
IF MAILED
IN THE
UNITED STATES

BUSINESS REPLY MAIL
FIRST CLASS MAIL PERMIT NO. 271 CARLSBAD, CA

POSTAGE WILL BE PAID BY ADDRESSEE

 Craftsman Book Company
6058 Corte del Cedro
P.O. Box 6500
Carlsbad, CA 92018-9974

NO POSTAGE
NECESSARY
IF MAILED
IN THE
UNITED STATES

BUSINESS REPLY MAIL
FIRST CLASS MAIL PERMIT NO. 271 CARLSBAD, CA

POSTAGE WILL BE PAID BY ADDRESSEE

 Craftsman Book Company
6058 Corte del Cedro
P.O. Box 6500
Carlsbad, CA 92018-9974

NO POSTAGE
NECESSARY
IF MAILED
IN THE
UNITED STATES